Fullerene Collision Reactions

Developments in Fullerene Science

Volume 5

Series Editor:

Tibor Braun, *Institute of Inorganic and Analytical Chemistry, L. Eötvös University, Budapest, Hungary*

Fullerene Collision Reactions

by

Eleanor E.B. Campbell

Göteborg University,
Göteborg, Sweden

SPRINGER-SCIENCE+BUSINESS MEDIA, B.V.

A C.I.P. Catalogue record for this book is available from the Library of Congress.

ISBN 978-90-481-6488-2 ISBN 978-1-4020-2524-2 (eBook)
DOI 10.1007/978-1-4020-2524-2

Printed on acid-free paper

Table of Contents

Chapter 1

INTRODUCTION

Fullerenes were first discovered in 1985 in an apparatus designed by Rick Smalley to produce atomic clusters of non-volatile elements [1]. The original aim of the experiment was to look for long chain HC_nN molecules that Harry Kroto thought could be formed under the laser vapourisation conditions of the experiment - conditions not unlike those in carbon stars [2]. The discovery of this new form of pure carbon, and the insight that led to the interpretation of the magic number features in the observed mass spectra in terms of hollow carbon molecules, led to the award of the Nobel Prize in Chemistry to Curl, Kroto and Smalley in 1996. Many interesting review articles and books have been written about the events surrounding the discovery [2-8]. However, the fullerenes would have remained a curiosity of interest only to a small community of chemical physicists studying the properties of atomic clusters if the breakthrough achieved by Krätschmer and Huffman in 1990 had not occurred [9]. While studying the optical properties of various kinds of graphitic soots, these scientists discovered a remarkably simple way of producing macroscopic amounts of fullerenes. In addition to the production, they also showed that it was possible to separate the stable fullerenes (in particular C_{60} and C_{70}) from the rest of the soot, to purify them by standard chemical methods and to produce sufficient material to confirm the original hollow "soccerball" suggestion for the structure of C_{60}.

Since then there have been very rapid developments in the field and fullerenes have made an impact in many different scientific disciplines. This monograph is just one in a series of books devoted to the properties of fullerenes. Fullerenes have perhaps made their most impressive impact, relative to the overall size of the research activities in a particular field worldwide, in the area of their birth: chemical physics and the closely related areas of atomic and molecular physics. The attraction of having a mass-selected cluster with the highest possible symmetry (I_h) that can easily be sublimed into the gas phase has inspired many groups with different experimental backgrounds to investigate the properties of these molecules. For many researchers from the atomic and molecular physics community this has opened up a whole new research area by providing a convenient and attractive model system for studying the behaviour of complex molecular systems with a large number of degrees of freedom. One nice example of this is a fullerene beam diffraction experiment showing quantum interference behaviour from C_{60} [10]. In many cases the experience gained from experiments with fullerenes is now beginning to be applied to even more complex biomolecular systems. From the theoretical point of view, the fascinating structure and high symmetry of buckminsterfullerene has also inspired many works on the structural, electronic and dynamical properties of the molecule. A subset of the fullerene activities within atomic, chemical and molecular physics: collision experiments with fullerenes, is the topic of this book.

The first two chapters following this introduction give a general introduction to the most common experimental techniques and theoretical models applied to the study of fullerene collisions. They supply background information that will be useful for understanding the discussions in later chapters. Chapters 4-10 then give overviews of the various sub-topics of the field. The aim is not to give a full comprehensive coverage of all work that has been carried out in any given topic but to pick out the most informative and illustrative examples and to point the reader towards more detailed presentations. On the whole, emphasis will be placed on collisions carried out under well-defined single-collision conditions. The topics range from reactive and thermal energy collisions (Chap 4) through collision induced dissociation (Chap 5), charge transfer (Chap 6) and electron impact studies (Chap 7). Chapters 8 and 9 are devoted to the more exotic members of the fullerene collision family: collisional production of endohedral fullerenes and fullerene-fullerene collisions. Finally, we finish in Chap. 10 with a discussion of the dynamical effects observed when fullerenes are collided from surfaces.

ACKNOWLEDGEMENTS

The completion of this book would not have been possible without the help of numerous colleagues who kindly sent pre- and reprints of their work and generously provided information and inspiration. I thank them all. I would also like to thank all undergraduate and graduate students and postdocs from my laboratory (at Freiburg University, the Max-Born-Institute in Berlin and Göteborg University) who have contributed so much to my own activities within the field of fullerene collisions over the years. Last but not least, my thanks go to my husband Mats who provided essential moral support and countless cups of coffee during the final stages of the manuscript preparation.

REFERENCES

1 H. W. Kroto, J. R. Heath, S. C. O'Brien, R. F. Curl and R. E. Smalley, Nature **318**, 162 (1985).
2 H. W. Kroto, W. Allaf and S. P. Balm, Chem. Revs. **91**, 1213 (1991).
3 J. Baggott, *Perfect Symmetry* (Oxford University Press, 1994).
4 R. F. Curl and R. E. Smalley, Science **242**, 1017 (1988).
5 R. F. Curl and R. E. Smalley, Scientific American **October**, 32 (1991).
6 H. W. Kroto, Science **242**, 1139 (1988).
7 R. E. Smalley, The Sciences **31**, 22 (1991).
8 G. Taubes, Science **253**, 1476 (1991).
9 W. Krätschmer, L. D. Lamb, K. Fostiropoulos and D. R. Huffmann, Nature **347**, 354 (1990).
10 M. Arndt, O. Nairz, J. Vos-Andreae, C. Keller, G. v. d. Zouw and A. Zeilinger, Nature **401**, 680 (1999).

Chapter 2

EXPERIMENTAL TECHNIQUES

Fullerenes have been very fruitful systems for studying collision dynamics. There are a number of reasons for this. One is the sheer beauty of the structure and the attraction of a spherical molecule. This has enticed a number of atomic and nuclear physics research groups into the area of molecular collisions and has opened up many new possibilities for applying the well-developed collision techniques of these well-established disciplines to study the dynamical behaviour of complex molecular systems. Fullerenes have provided an attractive and convenient stepping stone from the world of atomic and simple molecular collisions to large (bio)molecules. From a more pragmatic point of view, fullerenes are very easy to handle experimentally and this has of course also played a major role in the development of our understanding of these materials. Since molecular beams of fullerenes can be obtained with relative ease, as described below, this leaves more room for experimental complexity and creativity in the design of the collision experiments that can be carried out. The following chapter discusses the way fullerenes are used in collision experiments and explains the basic techniques that are employed. Examples are given of typical experimental set-ups covering a wide range in collision energies from thermal reactivity studies to MeV surface impact. Collision experiments require ultra-high vacuum ($< 10^{-10}$ mbar) or good high vacuum ($< 10^{-7}$ mbar) conditions since one wants to minimise the effects of collisions with background gas and is generally interested in the outcome of single collisions.

1. FULLERENE SOURCES

Fullerenes are prepared using the arc discharge method discovered by Krätschmer and Huffman in 1990 [1]. The soluble soot extract, which contains C_{60}, C_{70} and some higher fullerenes [2] is further purified by liquid chromatography (HPLC). Fullerenes are commercially available in varying degrees of purity from raw soluble soot material to "Gold grade" C_{60} with a nominal purity of 99.4 % (by weight). Many of the experiments that will be considered in this chapter and later in the book involve ionisation and mass selection of fullerene ions. For these purposes one does not require a highly purified sample but can use a mixture of extractable fullerenes since they will, in any case, be mass analysed before the collision. If neutral fullerenes are required – either as a molecular beam or as an effusive target, then the purity is more critical. Even highly purified commercial fullerene powder (> 99.4 % C_{60}) still contains a significant amount of impurities consisting of higher fullerenes and, particularly, solvent rest and possibly even compounds formed on passage through the HPLC purification column. The presence of these impurities also affects the vapour pressure and reproducibility of experimental results as will be discussed below. It is always advisable to carry out one or more sublimation cycles on commercial material, however this has only rarely been done in the experiments reported in the literature. Normally, the fullerene material will just be "out-gassed" at ca. 300 °C overnight before any experiments are carried out. The vapour pressure rises most steeply on the first increase in temperature (mainly solvent out-gassing) but on multiple heating and cooling cycles the achievable vapour pressure decreases. This effect cannot be solely due to solvent out-gassing. Many groups have also reported the formation of insoluble rest material in the oven after heating. The insoluble material still shows the presence of fullerenes if studied by e.g. laser desorption mass spectrometry and presumably has undergone a polymerisation reaction induced either by impurities in the powder and/or catalytic reactions involving trace elements from the oven walls.

1.1 Neutral Beams

The simplest way of producing a molecular beam of neutral fullerenes in a vacuum apparatus is to place the purified powder in a small oven with an aperture of a few mm and heat to a temperature of 350-500 °C. If a nozzle with a high aspect ratio is used with the oven then multiple collisions within the nozzle lead to a reasonably well-directed beam with the divergence decreasing with increasing nozzle length. The Kolodney group has developed an effusive source using ceramic material for a two-stage nozzle that can be heated to temperatures of around 2100 K [3]. This allows the temperature of the molecular beam to be varied over a relatively wide range that can be an important tool for determining the influence of thermal effects.

Due to the large number of vibrational degrees of freedom in C_{60} (174), the molecules in beams produced by simply heating the fullerene powder to a temperature of around 500 ^{o}C or higher have a relatively high amount of vibrational excitation which can affect the outcome of collision experiments. Fig. 2.1 plots the vibrational energy as a function of C_{60} temperature, calculated using the expression

$$E_{vib}(T) = \sum_i g_i h\nu_i \frac{\exp(-h\nu_i / k_B T)}{[1-\exp(-h\nu_i / k_B T)]} \qquad (2.1.1)$$

where g_i is the degeneracy of the vibrational mode, ν_i is the mode frequency, h is Planck's constant, k_B Boltzmann's constant and T the temperature. To a good approximation, the individual mode frequencies can be replaced by the average mode frequency (2.79x10^{13} Hz). The vibrational frequencies have been taken from Schettino et al. [4] where a complete vibrational assignment was performed using IR, Raman and neutron scattering data together with density functional calculations. The average of the vibrational mode frequencies reported in Schettino et al., as used to obtain Fig. 2.1, are tabulated in Table 2.1.

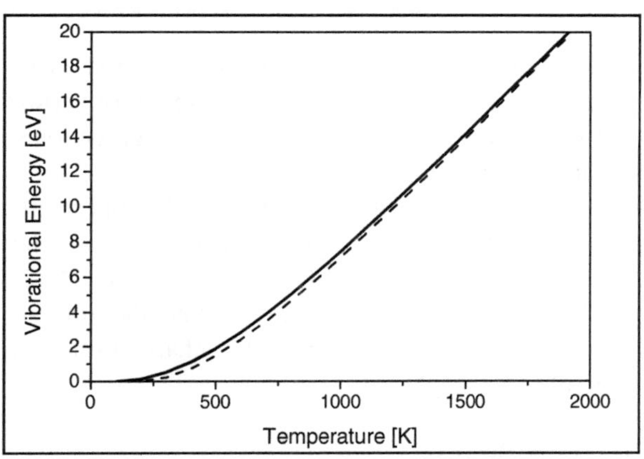

Figure 2-1. Relationship between vibrational energy and temperature for C_{60}, calculated using the vibrational frequencies in Table 2.1. The dashed line shows the calculated values using just the average vibrational frequency of 2.79x10^{13} Hz instead of the individual mode frequencies.

Table 2-1. Frequencies of normal modes of vibration of C_{60} in cm^{-1}. Average values taken from Schetttino et al.[4]. Degeneracies of the modes are given in brackets.

Symmetry	Frequency [cm^{-1}]	Symmetry	Frequency [cm^{-1}]
A_g (1)	491	A_u (1)	973
	1458	T_{1u} (3)	527
T_{1g} (3)	565		578
	850		1184
	1289		1439
T_{2g} (3)	535	T_{2u} (3)	355
	764		714
	805		1040
	1334		1187
G_g (4)	484		1565
	570	G_u (4)	344
	738		760
	1135		771
	1314		967
	1515		1321
H_g (5)	266		1415
	429	H_u (5)	403
	710		486
	768		669
	1095		740
	1254		1211
	1437		1342
	1570		1530

It is rather difficult to reduce the vibrational excitation energy in a beam of neutral fullerenes below the temperature of evaporation, however, this has been achieved by using a source based on a cluster aggregation source where the fullerenes are evaporated into a stream of He gas cooled to liquid nitrogen temperatures [5]. The fullerene vibrational temperature is then also liquid nitrogen temperature (or colder, depending on the expansion conditions). Such a source is presently being used in collisions between fullerenes and highly charged ions to investigate the role of vibrational excitation in determining the stability of highly-charged fullerenes produced in the collisions [6] (see Chapter 6).

1.2 Charged Beams

In the early nineties beams of charged (positive or negative) fullerenes used for collision experiments were often produced by laser desorption. This method was then mainly superseded (due to stability and cost arguments) by electron impact ionisation of an effusive beam of fullerenes produced from a small oven. Often, just a small tungsten filament biased at about 100 eV is enough to produce a relatively intense ion beam. The simplicity combined with the advantage of having either a continuous or pulsed ion source gives this method some clear advantages over the laser sources. Positive or negative ions can then be extracted from the source using standard techniques (some examples are given below). Many experiments that will be reported later in the book have used positively charged beams of fullerenes produced from electron cyclotron resonance (ECR) sources. Such sources were originally developed for producing highly charged ion beams and can be easily adapted for fullerenes [7]. An example of such a source is shown in Fig. 2.2. Beams of C_{60}^{q+} (q = 1-5) with intensities sufficiently high to carry out collision experiments can be routinely produced from ECR sources.

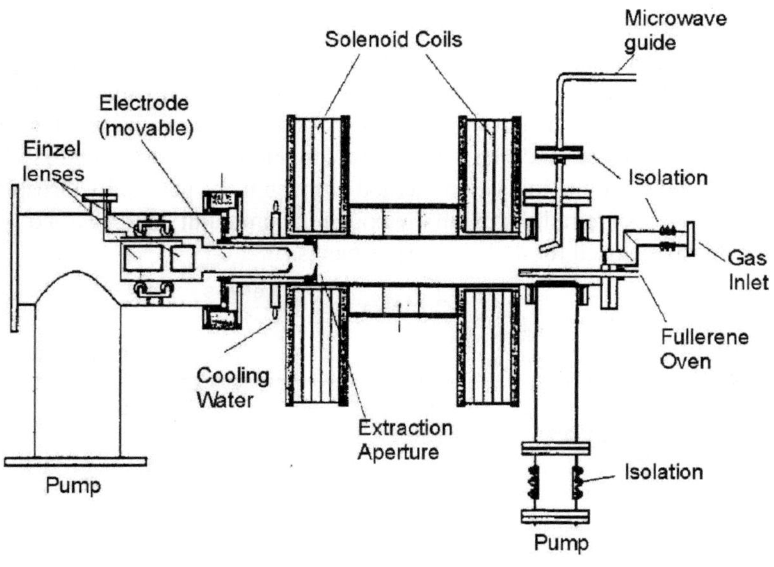

Figure 2-2. Schematic diagram of ECR source of multiply charged fullerenes, adapted from [8]

The big disadvantage of all the above-mentioned ion sources is the uncertainty in the internal excitation energy of the fullerene ions produced. An indication of the amount of excitation can be obtained from studying metastable decay of the fullerene ions on the microsecond time scale. This can be done, either in a conventional double sector mass spectrometer [9] or e.g. in a reflectron time-of-flight mass spectrometer [10]. The internal energy of the fullerene ions as they enter the field-free region of the mass spectrometer can be estimated by determining the amount of unimolecular decay (generally, evaporation of C_2 molecules from the cage) that occurs within a time window set by the parameters of the experiment. In this way the rate constant for unimolecular decay can be directly determined and, assuming the mechanism and energy for the unimolecular decay process are known, this can be used to estimate a vibrational temperature [11]. Unfortunately, the decay mechanism and dissociation energy for the fragmentation of C_{60} are still rather uncertain [12] (see Chap. 3.2.1), however, in any case it is safe to state that the fullerene ions emerging from electron impact ionisation (and laser desorption) sources are "hot" with internal energies on the order of tens of eV. Not only the magnitude but also the energy distribution is uncertain and is probably very broad (greater than thermal) for most of the ion sources used.

The situation is more promising for negatively charged fullerene ions. Recently, an electrospray ion source, shown in Fig. 2.3, has been developed that can produce fairly intense beams of C_{60}^- as long as a suitable solvent mixture is used [13]. The advantage of this kind of source is that the internal energy of the ions is much less than with conventional sources and is also better defined since the ionisation method is much more gentle. Electrospray fullerene sources are rather new in the collision community and have not yet been extensively used for collision experiments.

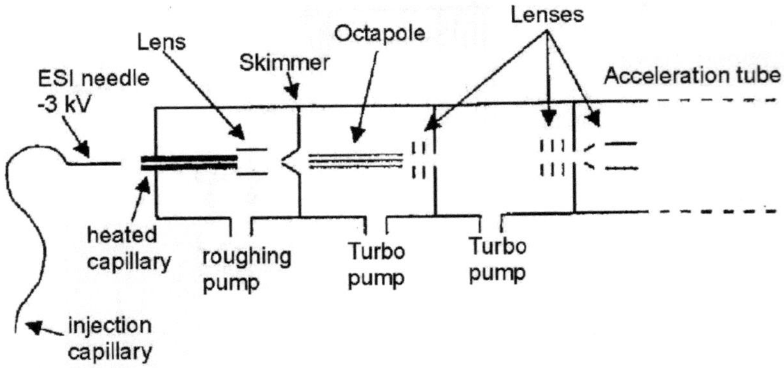

Figure 2-3. Schematic diagram of an electrospray ion source, adapted from [13].

1.3 Neutral targets

For many of the experiments reported later in the book, it is necessary to have a well-defined target of neutral fullerenes. In order to determine cross sections for reactions it is also necessary to know the density of the neutral fullerene target. The fullerene vapour pressure has been determined as a function of temperature by a number of groups. Unfortunately, the values are not in agreement and are in some cases out by orders of magnitude. Some of the available data is plotted in Fig. 2.4 with the fitted parameters and details of the measurements given in Table 2.2. The corresponding target density is plotted as a function of temperature in Fig. 2.5.

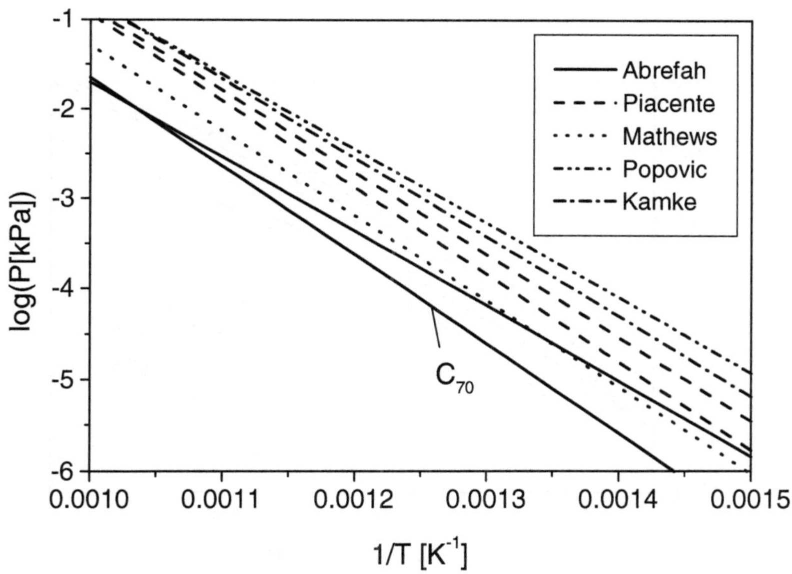

Figure 2-4. Log of vapour pressure as a function of inverse temperature. Straight lines: Data from Abrefah et al. [14] Their values for C_{70} are also shown; Dashed lines: Data from Piacente et al. [15], the upper line is obtained on the first sublimation, the lower line for a subsequent sublimation; Dotted line: from Mathews et al. [16]; Dash-dot from Jaensch and Kamke [17]; Dash-dot-dot from Popovic et al. [18].

Table 2-2. Comparison of Vapour Pressure Measurements.

		log(P/kPa) = A - B/(T/K)		
Method	Temperature Range /K	A	B	Reference
Knudsen Effusion	673-873	6.6 ± 0.1	8281 ± 873	[14]
Knudsen Effusion (C_{70})	703-873	8.2 ± 0.3	9848 ± 228	[14]
Torsion Effusion (1st)	730-990	8.28 ± 0.20	9154 ± 150	[15]
Torsion Effusion (2nd)	730-990	8.73 ± 0.25	9668 ± 200	[15]
Mass Spec.	600-800	8.18 ± 0.03	9475 ± 80	[16]
Mass Spec.	680-790	7.49 ± 0.14	8267 ± 120	[17]
Knudsen Evaporation	663-823	8.02 ± 0.19	8801 ± 145	[18]

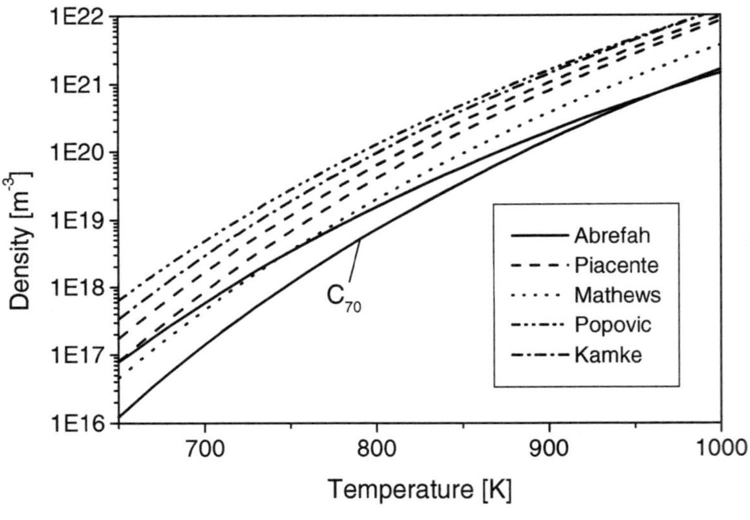

Figure 2.5. Target density as a function of temperature for C_{60} (with one set of data for C_{70}). Same sources as Fig. 2.4.

Although there is still some disagreement concerning the correct vapour pressure data, the collision community generally uses the values obtained by Abrefah et al. [14]. Normally, purification by sublimation is not carried out prior to collision experiments and the data obtained with the presence of residual solvents is most relevant for the determination of the target densities. A detailed discussion of the reasons for using the Abrefah data for analysing collision cross sections has been given by Schwartz et al. [19].

These authors base their choice of this data partly on their theoretical understanding of single-electron capture from C_{60} to highly charged ions (see Chapter 7). Their results are very consistent with the Abrefah data and not at all consistent with the Piacente, Popovic or Kamke values. The Piacente values have, however been shown to be consistent with UV-vis absorption measurements on pre-sublimed samples [20]. The Popovic and Kamke results may lie too high due to the complexity of the data analysis and the assumptions that have to be made concerning either ionisation cross sections and electron gain on the one hand [18], or details of the C_{60} - C_{60} interaction on the other [17]. All absolute cross section values discussed in later chapters have been obtained by using the Abrefah et al. vapour pressure results, unless otherwise stated.

1.3.1 Determination of Cross Sections

In all of the fullerene experiments which determine absolute reaction cross sections, a projectile beam is sent through a neutral target (normally a gas cell rather than a second beam) and the reaction products are determined as a function of collision energy (or scattering angle if differential cross sections are being determined). The intensity of the reaction product, $I_{product}$, is related to the cross-section for that reaction, σ, as follows:

$$I_{product} = I_0 n \sigma l \qquad (2.1.2)$$

where I_0 is the initial intensity of the projectile beam, n is the number density of the target gas and l is the length of the target gas cell. This expression holds in the target density range where single-collision events dominate and the target is not depleted. The absolute cross section for a given collision energy is determined by measuring the intensity of the reaction product as a function of the pressure (number density) in the target cell. For most experiments with rare gas targets this pressure can be determined directly with rather good accuracy. For fullerene targets, the temperature of the target cell, containing some fullerene powder, is varied and n determined with the help of the vapour pressure measurements discussed in the previous section. An example of typical data is shown in Fig. 2.6. The gradient in the linear, low-pressure range of the data, where single-collisions dominate, yields the cross-section. Deviation from straight-line behaviour occurs due to the increasing importance of multiple collisions in the target cell. Such pressure/temperature dependent measurements can be very time consuming and often this will be carried out at just two or three values of the collision energy. Relative determinations will then be carried out for the other collision energies at constant pressure/temperature.

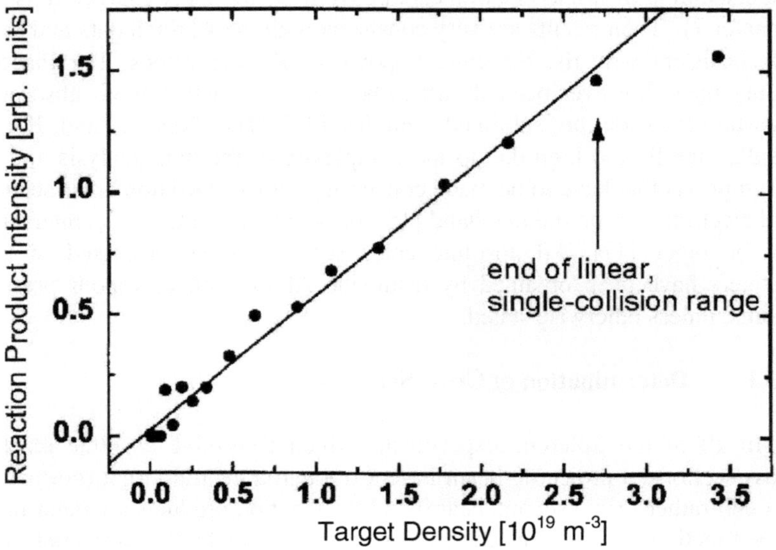

Figure 2.6. Example of the determination of collision cross-sections. The reaction product ion intensity (in this case C_{120}^+ from C_{60}^+ + C_{60} collisions) is determined as a function of target gas number density. The cross section is obtained from a least squares fit to the data in the low target density range where single-collision conditions are operating. Adapted from [21].

The detection efficiency of the projectile beam (normally positive ions) should also be considered, in particular if a wide range of masses is being detected and/or collisions over a wide collision energy range are being investigated. Microchannel plates are often the ion detectors of choice. The detection efficiency in this case depends strongly on the velocity of the species being detected. Heavy fullerene ions will be detected with much less efficiency than faster, smaller fragment ions. This has to be investigated for each individual experimental set-up (different microchannel plates can show quite different behaviour) and the raw data corrected accordingly. An example of such a procedure is given by Campbell et al. [22] and applied to fullerene projectile collision experiments in [23]. An example of the effect on a fragmentation mass spectrum of correcting for the mass (or velocity) dependence of the detection efficiency is shown in Fig. 2.7. The problem becomes much less severe when the fullerene ions have velocities beyond ca. 5×10^4 ms^{-1} (or collision energies beyond ca. 10 keV for C_{60}^+) where the detection efficiencies converge.

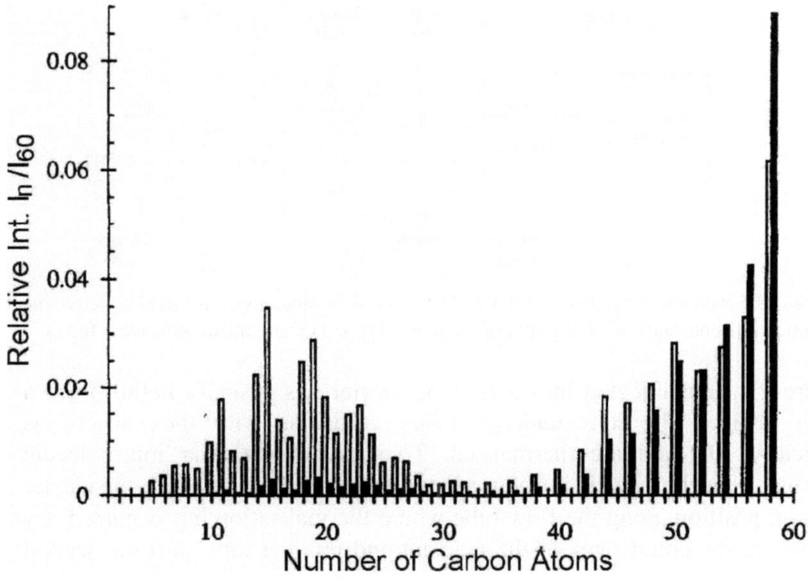

Fig. 2.7. Relative fragment intensities as measured directly in C_{60}^+ + Ar experiments at a centre of mass collision energy of 85 eV (open bars) compared to the intensity corrected for different detection efficiences of the carbon clusters (filled bars). From [23].

2. EXAMPLES OF COLLISION EXPERIMENTS

2.1 Thermal Collision Energies

The majority of reactive collision experiments carried out at thermal collision energies have used the technique known as SIFT: Selected Ion Flow Tube. This technique is a fast flow tube / ion swarm method for the study of the reactions of ions (positive or negative) with atoms and molecules under thermalised conditions over a wide range of temperatures. It has been extensively used to study ion-molecule kinetics. In contrast to most of the techniques used to investigate fullerene collisions, this does not study the products of individual collisions but is used to determine reaction rate constants. In the SIFT apparatus the ions are created in an ion source which is external to the flow tube. In order to study the reactions of fullerene ions, C_{60}^{q+}, the ions were produced by electron impact ionisation of vaporised material [24]. The ions are then extracted from the ion source, selected according to their mass-to-charge ratio using a quadrupole mass filter, see

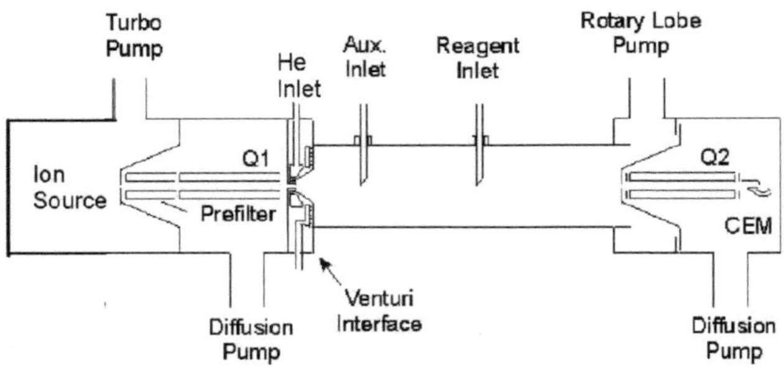

Figure 2-8. Schematic diagram of a SIFT (selected ion flow tube) apparatus used to determine reaction rate constants involving fullerene cations. Q1 and Q2 are quadrupole mass filters.

Figure 2.8, and injected into a flowing carrier gas (usually helium) via a small orifice. The ions undergo many collisions with the carrier gas (typically 10^5) and are thermalised. To study a particular ion-molecule reaction, a reactant gas is introduced at a measured flow rate into the carrier gas at a position along the flow tube where thermalisation has occurred. By measuring the count rates of the reactant and product ions with the second quadrupole mass spectrometer and relating them to the number density of the reactant gas it is possible to determine the rate constant for that particular reaction.

The reaction rate coefficient determined in such flow tube experiments (in units of $cm^3mol^{-1}s^{-1}$) can be related to the cross section for the given reaction as $k(\mathrm{v}) = \mathrm{v}\sigma(\mathrm{v})$ with v the relative collision velocity. Assuming that the velocity distribution in the experiments is thermal with a temperature, T then the normalised distribution for the relative velocity in the centre-of-mass frame of reference is

$$f(\mathrm{v})d\mathrm{v} = 4\pi \mathrm{v}^2 \left(\mu / 2\pi k_B T\right)^{3/2} \exp\left(-\mu \mathrm{v}^2 / 2k_B T\right) dv \qquad (2.2.1)$$

where μ is the reduced mass. The rate constant at temperature T is then given by

$$k(T) = \int f(\mathrm{v})\sigma(\mathrm{v})d\mathrm{v}. \qquad (2.2.2)$$

If we assume that the cross section depends only on the collision velocity or relative collision energy, E_{CM} (and not on the details of the initial and final states) and make the further rather extreme assumption that the collision cross section can be given by the line-of-centres model, discussed in Chapter 9 then

$$\sigma = \pi d^2 (1 - E_0 / E_{CM}) \text{ for } E_{CM} \geq E_0$$
$$= 0 \text{ for } E_{CM} < E_0 \qquad (2.2.3)$$

with E_0 the threshold energy for reaction. Eq. (2.2.2) then becomes [25]

$$k(T) = \pi d^2 (8k_B T / \pi\mu)^{1/2} \exp(-E_0 / k_B T). \qquad (2.2.4)$$

This is just the standard "collision theory" rate constant with an Arrhenius-like temperature dependence. Note that a number of simplifying assumptions have gone into deriving this expression and normally the pre-exponential factor determined in experiments is not in good agreement with the simple derivation. Also, it is quite common in SIFT experiments of ion-atom or ion-molecule collisions to apply a small electric field to "pull" the ions through the neutral target. In this case the relative velocity distribution is different from eq. (2.2.1).

2.2 Hyperthermal collision experiments

Hyperthermal collision experiments can be divided into two different energy regimes: low and high. "Low" in this case refers to collisions where the collision velocity is lower than the orbital velocity of the electrons. In this case, the main excitation channel is expected to be vibrational excitation. At "high" collision energies the excitation is dominated by electronic transitions. This is described in more detail in the following Chapter. Most of the studies in the literature fall into the "low" energy category with collision energies on the order of a few keV. The same kind of experimental set-up is used for both categories of hyperthermal collision. For these collisions fullerenes can either be the projectile ions (positively or negatively charged) colliding with a neutral gas target or the neutral targets in collisions with singly or multiply charged atomic ions or both projectile and target. Figure 2.9 shows an example of an experimental set-up used to study low energy collisions of fullerene ions with neutral targets, including neutral fullerenes [21]. This is one of the few set-ups used to study fullerene collisions where it is also possible to determine the angular distributions of the collision products. The fullerene ions are produced by electron impact ionisation and are then accelerated with a pulsed electric field into a small time-of-flight mass spectrometer. The ions are focused in time according to their mass to charge ratio [26] and can be mass-selected by briefly switching off an applied electric field when the ions of choice are present at the "mass selector". The mass-selected ions are then steered and focused into a small target chamber where the collisions with the neutral target gas take place under single-collision conditions. The collision energy is controlled either by changing the acceleration voltage in the source region or by changing the potential of the target chamber and can be varied from 50 eV to 5 keV in the laboratory frame of reference. The energy resolution is typically on the order of 5% of the laboratory collision energy. Charged products from the collision are detected in a reflectron time-of-flight mass spectrometer [27] with an additional retarding field energy detector. The reflectron detector assembly can also be rotated around the target using a specially constructed target oven with a horizontal exit slit. This has been used to determine differential cross sections in fullerene cation - neutral fullerene collisions [28].

The apparatus has also been used to study positively charged products of hyperthermal collisions of neutral fullerenes with neutral gas targets [29]. In this case, a laser desorption source was used to produce an intense beam consisting of both positively and negatively charged fullerenes. The negatively charged fullerenes were accelerated to the desired laboratory collision energy and mass-selected. After mass selection a significant proportion of the negatively charged ions undergoes thermal electron emission due to the high internal vibrational excitation they receive from the laser excitation. Just before entering the collision cell, the remaining negative ions are removed from the beam with an electric field and only the accelerated neutral fullerenes enter the target chamber.

Instead of time-of-flight mass spectrometers, it is quite common to use electric and magnetic field sector mass spectrometers for ion selection before and after collision. This gives the advantage of higher mass and energy resolution but they tend to be less flexible and it is more difficult to scan parameters such as the collision energy over a large range. One nice example of a set-up based on electric and magnetic sector mass spectrometers is that developed by the Schwarz group [30] and used mainly to study collision induced dissociation of molecular cations (including fullerenes). This is the apparatus that was used in experiments that first showed the possibility of collisionally capturing an atom inside the fullerene cage to produce endohedral fullerenes (see Chapter 8). A schematic diagram of the apparatus is shown in Fig. 2.10. It is optimised to work at a laboratory collision energy of 5 keV. The big advantage of the apparatus is the possibility of carrying out successive collisions followed by mass/energy selection. With the constellation shown in Fig. 2.10 it is e.g. possible to carry out the following

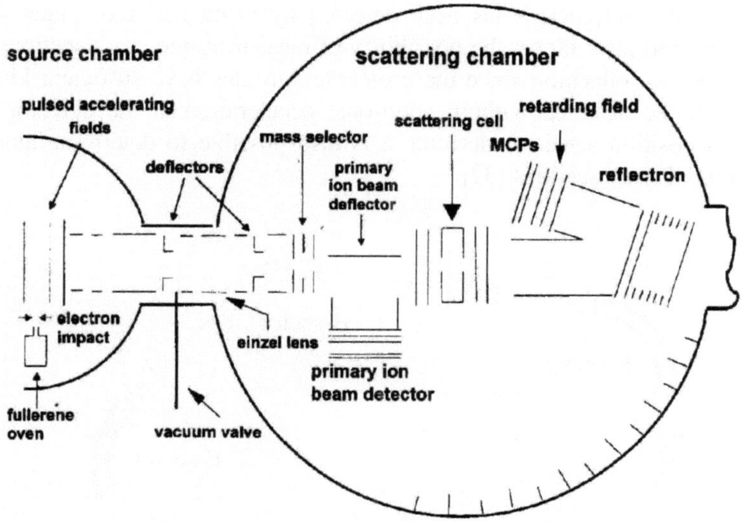

Figure 2.9. Schematic diagram of SOCCER apparatus (Scattering of Carbon Clusters Energy Resolved) used to study collisions between positively charged fullerene ions and neutral fullerene targets. From [21].

experiment. Fullerene ions are produced by electron impact ionisation in the source region and then selected according to their mass/charge ratio. They then collide with a neutral target gas, e.g. He and a small fraction of the fullerene ions capture the He atom inside their cage. These ions are then selected by sector 2, collide in a second target chamber where they can undergo a neutralisation reaction. After this collision chamber any residual positive ions are removed from the beam by applying an electric field and the fast neutral molecules enter the third collision chamber where they can again undergo a charge transfer reaction to emerge as cations. These undergo a final mass-selection before being detected. Such a complicated and challenging series of collisions was carried out as described [31] and was one of the early strong indications of the endohedral nature of the ions produced in the first collisions.

Many hyperthermal energy collisions have been carried out using neutral fullerenes as the target. Often, highly charged ions are used as the projectiles and the cross sections for electron transfer from the fullerene to the projectile are studied (see Chapter 7). Examples of these kinds of collision experiments are shown in Figs. 2.11 and 2.12. In Fig. 2.11 the charge state of the projectile ion after collision as well as any kinetic energy gain or loss is determined as a function of collision energy and initial charge state. After collision the projectile ions are selected according to their energy and charge state by means of a 180° cylindrical energy analyser. Alternatively, the final charge state and energy has been detected by using deflector plates. The latter method also allows the possibility of measuring the cross sections for projectile neutralisation since the projectile particles have sufficient kinetic energy to be detected without additional acceleration at the detector. By using a position sensitive detector it is also possible to determine angular differential cross sections [32].

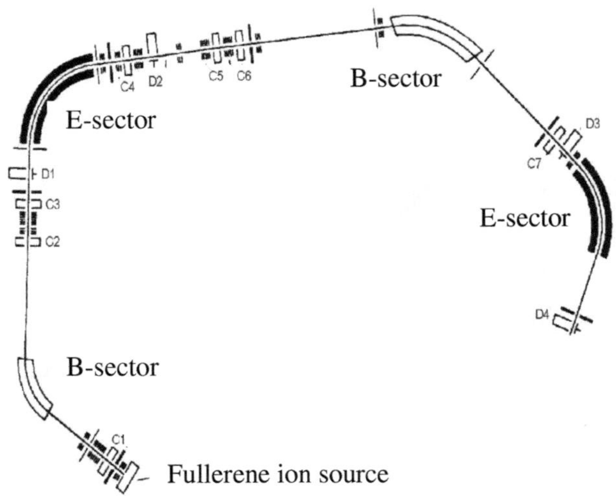

Figure 2-10. Schematic diagram of Schwarz experiment. C1-7, collision cells; D1-4, ion detectors. Adapted from [33].

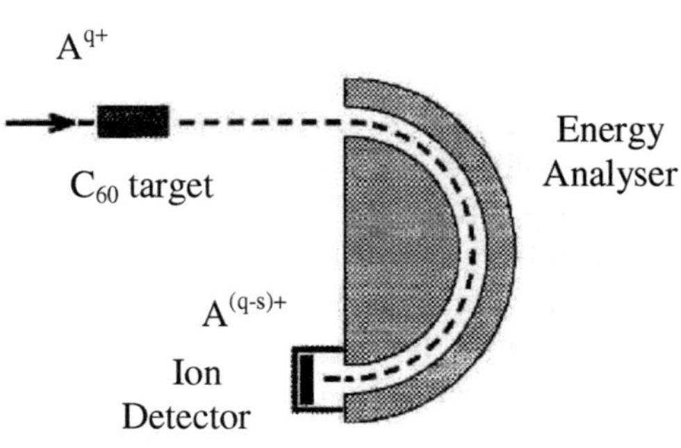

Figure 2.11. Apparatus used to measure total electron capture cross-sections in collisions between multiply charged ions, A^{q+}, and C_{60}. Adapted from [19]

In the second set-up (Fig. 2.12), the ionised target fullerenes are detected using a linear time-of-flight mass spectrometer. This technique has the advantage of simultaneously detecting the parent target ions as well as their fragment ions thus giving an indication of the amount of energy transferred

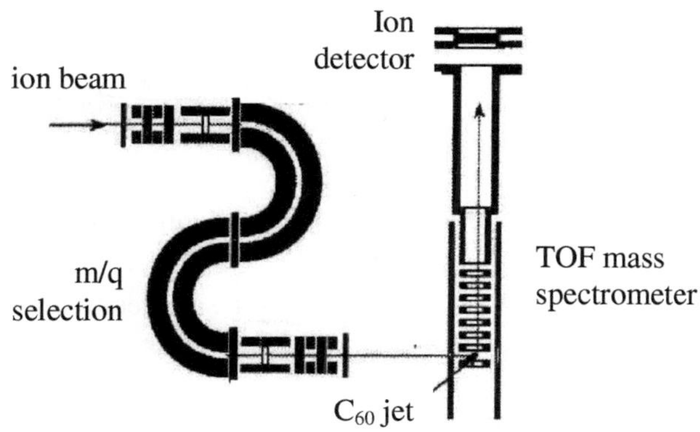

to internal degrees of freedom in the collisions.

Figure 2-12. Apparatus for studying charge transfer cross sections in collisions between multiply charged ions (including C_{60}^{q+}) and neutral C_{60}. Product ions and their fragments are detected with a time-of-flight mass spectrometer. Adapted from [34].

2.3 Coincidence experiments

In all of the experiments discussed so far it is only the positively charged products of the collisions that are detected. This is of course a disadvantage when it comes to determining collision mechanisms and obtaining information on energy transfer etc. since much information is "hidden" in the neutral fragment distributions or emitted electron kinetic energies, for example. Much more detailed information can be obtained from coincidence experiments and a few coincidence measurements have been carried out in recent years to study the products of fullerene collisions.

An early example is the set-up used by Vandenbosch and co-workers, shown in Fig. 2.13 [35]. In this experiment, an energetic beam of mass-selected negatively charged C_{60} is produced and collides with a light target atom or molecule (e.g. H_2). Due to the kinematics of the collision, the fullerene fragments continue in a narrow cone about the original projectile beam axis. The positively charged fragment ions produced in the collision are then detected in coincidence. This allows the size distribution of the partner fragment ions to a particular mass fragment to be determined. The more energetic heavier fragments pass through a hole in the first detector and enter the second detector. The heavy fragments detected in this way are then measured in coincidence with the different light fragments detected with detector 1. This experiment provided evidence for a binary break-up of the fullerene cage followed by sequential evaporation of C_2 molecules from the hot fragment ions produced in this way (see Chapter 5).

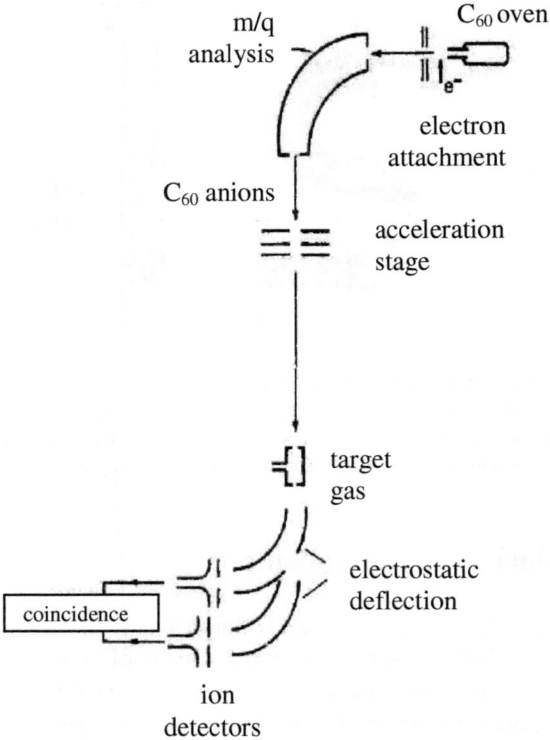

Figure 2.13. Apparatus used to measure coincidences between fragment ions produced in energetic collisions between C_{60} anions and light target gases. Adapted from [35].

The apparatus shown in Fig. 2.14 is quite different from that of Fig. 2.13 but is used to obtain similar information [36]. In this case, fast protons collide with a neutral fullerene target. Electrons produced in the energetic collisions are used as the start signal for a time and position sensitive detector. In this way all positively charged fragments are detected in coincidence with one of the emitted electrons.

The kind of set-ups in Figs. 2.11 and 2.12 have been combined to determine the target charge and fragmentation distributions in coincidence with a given charge-exchanged projectile ion, Fig. 2.15. Information on the kinetic energy distributions of the electrons emitted in collisions between highly charged ions and neutral fullerenes has been obtained by introducing an electron energy analyser and measuring the energy distributions in coincidence with time-of-flight mass spectra (similar to the set-up in Fig. 2.15) [37].

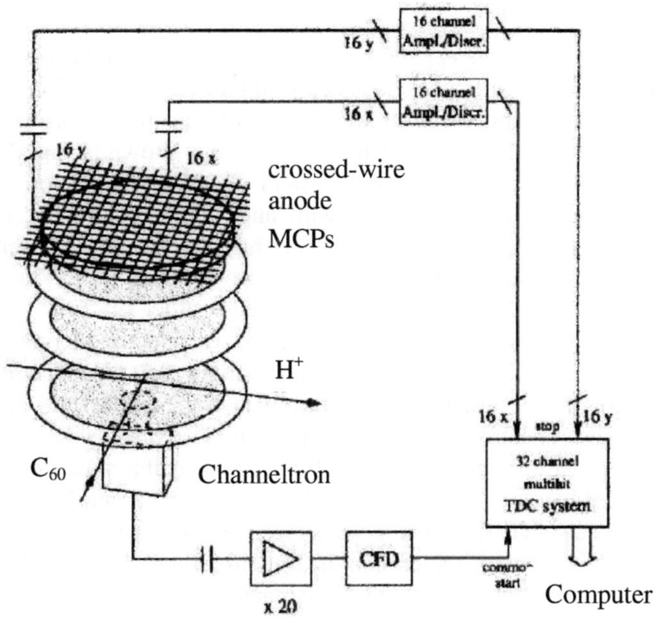

Figure 2.14. Apparatus for determining coincidences between fragment ions produced in collisions between energetic protons and neutral C_{60}. Adapted from [36].

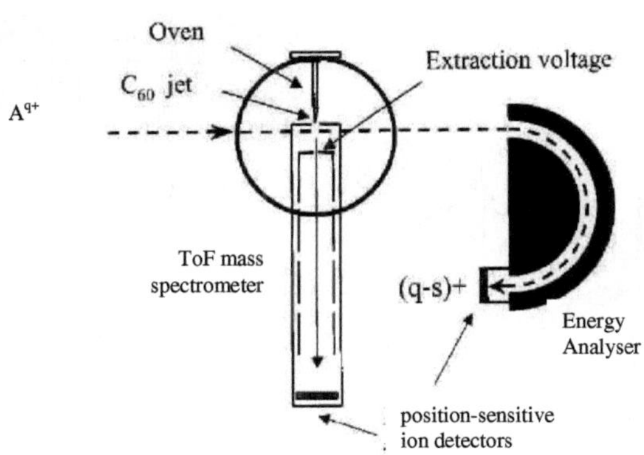

Figure 2.15. Apparatus to measure the energy distributions of the charge-exchanged projectile ions in coincidence with the fullerene fragmentation spectrum. Adapted from [37].

2.4 Electron impact

The collision energy dependence of electron impact ionisation and fragmentation of fullerenes has been studied extensively. Many of the results involving energetic electron collisions with neutral fullerenes have been obtained in the Märk group using the apparatus shown in Fig. 2.16 [9]. This is a double-focusing two-sector field mass spectrometer. An effusive fullerene beam interacts at 90° with an electron beam. For cross section measurements the electron current is reduced to ca. 10 μA to ensure single-collision conditions. Higher currents can be used to produce significant amounts of multiply-charged ions in higher order reactions. The electrons are guided by a weak magnetic field and have kinetic energies in the range 1-1000 eV with an energy spread of about 0.5 eV. The positive ions produced in the collision are extracted at right angles and enter the two-sector mass spectrometer through the slit S1. Metastable decay reactions occurring within the second field-free region of the mass spectrometer can be studied by using the so-called MIKE scan technique [38]. Measurement of the metastable fragment peak shape, obtained by scanning the electric field in the second sector can be directly related to the kinetic energy released during the metastable fragmentation. This, in turn, can provide information on dissociation energies although it is dependent on the model assumed for the fragmentation mechanism. This has led to problems concerning the determination of the dissociation energies for fullerenes (see

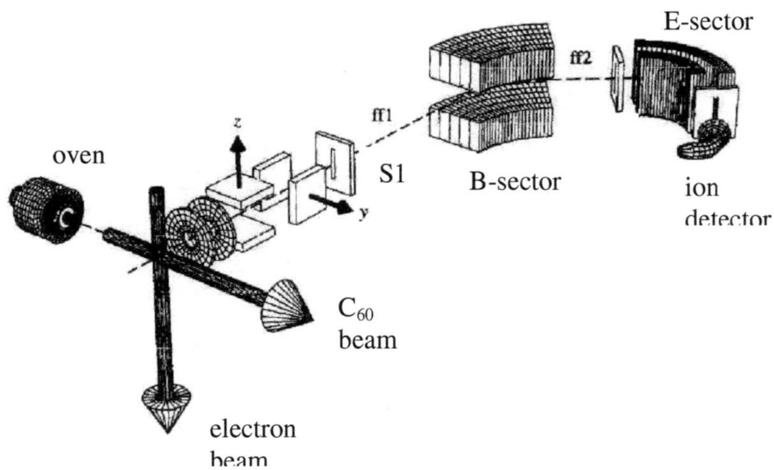

Figure 2.16. Two-sector mass spectrometer used extensively to study the metastable fragmentation of fullerene ions produced by energetic electron impact. Fullerene ions are mass-selected by the magnetic sector and the fragmentation that occurs in the second field-free region (ff2) is studied by scanning the electric field in the second sector of the machine, Adapted from [39].

Chap. 3.2.1). Recently the apparatus has been upgraded to a three-sector instrument, which has led to a substantial improvement in the achievable resolution [39].

The same apparatus has also been used to study electron collisions with positively charged fullerene ions by inserting an intense electron beam between the two sectors [40]. However, most of the literature results concerning cross sections for electron impact ionisation and fragmentation of fullerene ions have been obtained with the apparatus developed by the Salzborn group and shown in Fig. 2.17 [41]. The fullerene ions were produced in an ECR ion source (see 2.1.2) and then crossed an intense electron beam with energy in the range 10-1000 eV. After the interaction the product ions were separated by a 90° magnet and detected with a single-particle detector. Absolute cross sections were obtained by using the dynamic crossed beam method where the electron beam is moved through the ion beam with simultaneous registration of the parent and product ion beam intensities [42].

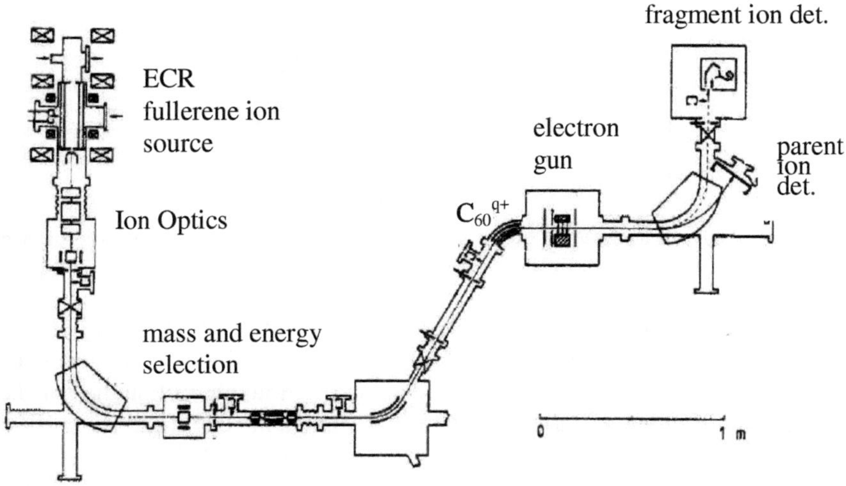

Figure 2-17. Experimental set-up for determining absolute cross-sections in ion-electron collisions. Fullerene ions are produced in an ECR source, mass and energy selected and then collide with a high power electron gun. Both the fragment ion distribution and the parent ion beam intensity is recorded. Adapted from [41].

2.5 Surface Collisions

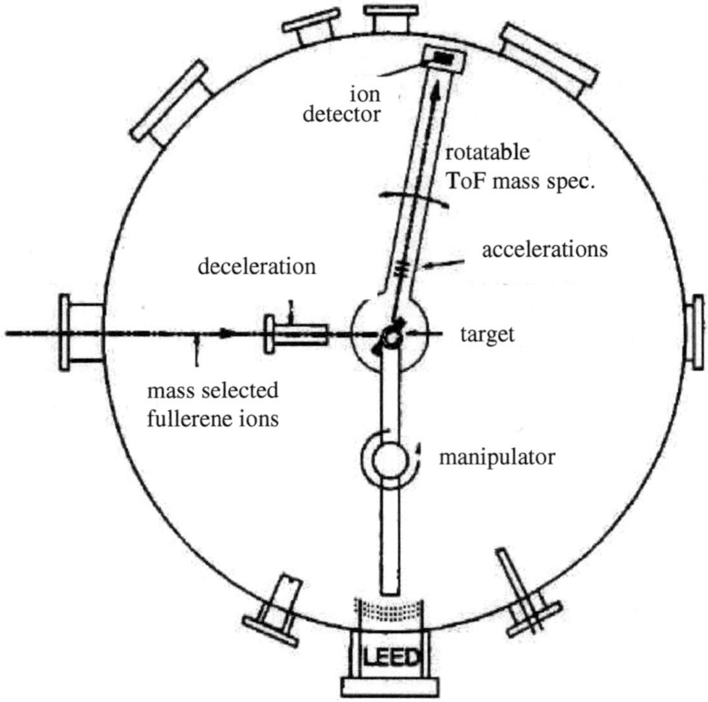

Figure 2.18. Apparatus used for studying the collisions of fullerene ions on surfaces. Adapted from [43].

Fullerenes colliding with surfaces at relatively high impact energies of a few hundred eV and higher were shown to be very resilient with a tendency to "bounce" away from the surface. This led to a number of studies of the dynamics of fullerene surface collisions, discussed in detail in Chapter 10. A typical surface scattering apparatus is shown in Fig. 2.18 [44]. The set-up is quite similar to the gas phase scattering experiment shown in Fig. 2.9. In the surface scattering experiment, laser desorption is used as the fullerene ion source. The fullerenes are directly desorbed into a pulsed expansion of a rare gas. Fullerene cations, anions and neutrals are cooled and carried by the rare gas pulse through a skimmer and into an acceleration region. The ions are accelerated perpendicular to the direction of expansion of the gas pulse and mass selected with a pulsed electric field. A pulsed deceleration lens is used to decelerate the fullerene ions to the desired collision energy thus allowing the surface region to remain field-free. After scattering from the surface the ions are pulsed accelerated and detected with a rotatable time-of-flight mass spectrometer. The velocity distribution of the scattered ions can be mapped out by changing the time delay between ion impact on the surface and pulsing on of the extraction acceleration field. The scattering dynamics can be strongly influenced by impurities on the target surface. For this reason it is essential to clean the surface by sputtering and to control the surface quality by LEED and Auger spectroscopy.

Fullerenes have also been used to study the effects of implanting high energy (MeV) clusters into solids [45]. For these studies the fullerene ions were produced in a tandem accelerator where a Cs ion gun is used to bombard a C_{60} target in order to produce negatively charged ions with a few keV energy. These ions are then accelerated to MeV energies and collide with N_2 gas to produce positive ions that are then further accelerated in the second part of the tandem.

REFERENCES

1 W. Krätschmer, L. D. Lamb, K. Fostiropoulos and D. R. Huffmann, Nature **347**, 354 (1990).

2 G. Ulmer, E. E. B. Campbell, R. Kuhnle, H.-G. Busmann and I. V. Hertel, Chem. Phys. Lett. **182**, 114 (1991).

3 A. Budrevich, B. Tsipinyuk and E. Kolodney, J. Phys. B **29**, 4965 (1996).

4 V. Schettino, P. R. Salvi, R. Bini and G. Cardini, J. Chem. Phys. **101**, 11079 (1994).

5 K. Hansen, R. Müller, P. Brockhaus, E. E. B. Campbell and I. V. Hertel, Z. Phys. D **42**, 153 (1997).

6 T. Bergen, X. Biquard, A. Brenac, F. Chandezon, B. A. Huber, D. Jalabert, H. Lebius, *et al.*, Rev. Sci. Instrum. **70**, 3244.

7 S. Biri, A. Valek, L. Kenez, A. Janossy and A. Kitagawa, Rev. Sci. Instrum. **73**, 881 (2002).

8 M. Liehr, M. Schlapp, R. Trassl, G. Hofmann, M. Stenke, R. Völpel and E. Salzborn, Nucl. Instrum. Methods Phys. Res. B **79**, 697 (1993).

9 S. Matt, P. Scheier, A. Stamatovic, H. Deutsch, K. Becker and T. D. Märk, Philos. Trans. R. Soc. Lond. A **357**, 1201 (1999).

10 E. E. B. Campbell, G. Ulmer, H.-G. Busmann and I. V. Hertel, Chem. Phys. Lett. **175**, 505 (1990).

11 H. Hohmann, R. Ehlich, S. Furrer, O. Kittelmann, J. Ringling and E. E. B. Campbell, Z. Phys. D **33**, 143 (1995).

12 C. Lifshitz, Int. J. Mass Spec. **198**, 1 (2000).

13 O. V. Boltalina, P. Hvelplund, T. J. D. Jørgensen, M. C. Larsen, M. O. Larsson and D. A. Sharoitchenko, Phys. Rev. A **62**, 023202 (2000).

14 J. Abrefah, D. R. Olander, M. Balooch and J. Siekhaus, Appl. Phys. Lett. **60**, 1313 (1992).

15 V. Piacente, G. Gigli, P. Scardala, A. Guistini and D. Ferro, J. Phys. Chem. **99**, 14052 (1995).
16 C. K. Mathews, M. Sai Baba, T. S. Lakshmi Narasimhan, R. Balasubramanian, N. Sivaraman, T. G. Srinivasan and P. R. Vasudeva Rao, J. Phys. Chem. **96**, 3566 (1992).
17 R. Jaensch and W. Kamke, Mol. Materials., Proc. of IWFAC 1999 (2000).
18 A. Popovic, G. Drazic and J. Marsel, Rapid Commun. Mass Spectrom. **8**, 985 (1994).
19 S. H. Schwartz, A. Fardi, K. Haghighat, A. Langereis, H. T. Schmidt and H. Cederquist, Phys. Rev. A **63**, 013201 (2001).
20 P. F. Coheur, M. Carleer and R. Colin, J. Phys. B: At. Mol. Opt. Phys. **29**, 4987 (1996).
21 F. Rohmund, A. V. Glotov, K. Hansen and E. E. B. Campbell, J. Phys. B. **29**, 5143 (1996).
22 E. E. B. Campbell, G. Ulmer, B. Hasselberger, H.-G. Busmann and I. V. Hertel, J. Chem. Phys. **93**, 6900 (1990).
23 D. K. Bohme, Int. Rev. Phys. Chem. **13**, 163 (1994).
24 R. D. Levine and R. B. Bernstein, *Molecular Reaction Dynamics and Chemical Reactivity* (Oxford University Press, New York, Oxford, 1987).
25 W. C. Wiley and I. H. McLaren, Rev. Sci. Instrum. **26**, 1150 (1955).
26 B. A. Mamyrin, V. J. Karatajev, D. V. Shmikk and V. A. Zagulin, Sov. Phys. JETP **37**, 45 (1973).
27 A. V. Glotov and E. E. B. Campbell, Phys. Rev. A **62**, 033202 (2000).
28 R. Ehlich, H. Sprang, M. Westerburg and E. E. B. Campbell, J. Chem. Phys. **108**, 9390 (1998).
29 H. Schwarz, Angew. Chem. Int. Ed. Engl. **32**, 1412 (1993).
30 T. Weiske, T. Wong, W. Krätschmer, J. K. Terlouw and H. Schwarz, Angew. Chem. Int. Ed. Engl. **31**, 183 (1992).
31 H. Cederquist, A. Fardi, K. Haghighat, A. Langereis, H. T. Schmidt, S. H. Schwartz, J. C. Levin, *et al.*, Phys. Rev. A **61**, 022712 (2000).
32 H. Schwarz, T. Weiske, D. K. Böhme and J. Hrusák, in *Buckminsterfullerenes*, edited by W. E. Billups and M. A. Ciufiolini (VCH Publishers, New York, 1993), p. 257.
33 H. Cederquist, P. Hvelplund, H. Lebius, H. T. Schmidt, S. Tomita and B. A. Huber, Phys. Rev. A **63**, 025201 (2001).
34 R. Vandenbosch, B. P. Henry, C. Cooper, M. L. Gardel, J. F. Liang and D. I. Will, Phys. Rev. Lett. **81**, 1821 (1998).
35 A. Reinköster, U. Werner, N. M. Kabachnik and H. O. Lutz, Phys. Rev. A. **64**, 023201 (2001).
36 J. Jensen, A. Langereis, H. Cederquist, A. Fardi, H. T. Schmidt, S. H. Schwartz and H. Zettergren, MSL Annual Report (2000).
37 R. Cooks, J. H. Beynon, R. M. Caprioli and G. R. Lester, *Metastable Ions* (Elsevier, Amsterdam, 1973).
38 T. Fiegele, V. Grill, S. Matt, M. Lezius, G. Hanel, M. Probst, P. Scheier, *et al.*, Vacuum **63**, 561 (2001).
39 S. Matt, O. Echt, T. Rauth, B. Duenser, M. Lezius, A. Stamatovic, P. Scheier, *et al.*, Z. Phys. D **40**, 389 (1997).
40 K. Tinschert, A. Muller, G. Hofmann, K. Huber, R. Becker, D. C. Gregory and E. Salzborn, J. Phys. B **22**, 531 (1989).
41 A. Mueller, K. Tinschert, C. Achenbach, E. Salzborn and R. Becker, Nucl. Instrum. Methods Phys. Res. B **10/11**, 204 (1985).
42 R. D. Beck, P. Weis, G. Bräuchle and J. Rockenberger, Rev. Sci. Instrum. **66**, 4188 (1995).
43 P. Weis, J. Rockenberger, R. D. Beck and M. M. Kappes, J. Chem. Phys. **104** (1996).
44 A. Dunlop, G. Jaskierowicz, J. Jensen and S. Della-Negra, Nucl. Inst. Meth. in Phys. Res. B **132**, 93 (1997).

Chapter 3

THEORETICAL MODELS

In this chapter we will review some basic collision dynamics and convenient frames of reference. More detailed discussions can be found in text books dealing with atomic and molecular collisions e.g. [1-4]. We will also discuss some of the theoretical models used to describe the general dynamical behaviour of fullerenes. Theoretical considerations that are specific to a certain sub-set of collision experiments will be covered in the relevant chapter.

1. FRAMES OF REFERENCE AND CLASSICAL MECHANICS OF COLLISIONS

Two frames of reference are normally used to describe collision experiments: the laboratory and the centre-of-mass frames. The laboratory frame of reference is the one most convenient for experimentalists. It describes the experimental results in terms of the kinetic energy or velocity of the projectile ion beam and, if appropriate, the scattering angle of the projectile beam as directly measured in the laboratory. This is a convenient frame of reference to use at high collision energies where the velocity of the projectile with respect to the electron orbital velocities in the target atom or molecule becomes a relevant factor (see below). However, at low collision energies it is normally more convenient to think in terms of the centre-of-mass reference frame. The force between the particles depends only on the relative distance between them. For this reason it is of advantage to describe the collision in terms of the relative separation of the colliding partners and the motion of their centre-of-mass rather than the motion of each individual partner. If no external forces are acting then the motion of the centre-of-mass is unchanged during the collision. It can therefore be extracted from the total energy and we only need to consider the remainder, which is the kinetic energy of the relative motion. The relative energy of the collision partners is important since this determines the maximum energy available for conversion to internal excitation in the collision. In a fully inelastic collision the entire centre-of-mass collision energy is converted to internal excitation and the relative velocity of the collision partners after collision is zero (e.g. in the experiments discussed in Chapters 8 and 9). The velocity of collision partner i in the laboratory frame can be written as

$$\mathbf{v}_i = \mathbf{v}_{cm} + \mathbf{w}_i \tag{3.1.1}$$

where $\mathbf{v}_{cm}$ is the velocity of the centre of mass

$$\mathbf{v}_{cm} = \frac{(m_1\mathbf{v}_1 + m_2\mathbf{v}_2)}{(m_1 + m_2)} \tag{3.1.2}$$

and $\mathbf{w}_i$ is the velocity of particle i with respect to the centre-of-mass. Conservation of linear momentum implies that

$$m_1\mathbf{w}_1 + m_2\mathbf{w}_2 = 0 \tag{3.1.3}$$

where m_1 and m_2 are the masses of the collision partners. The relative velocity is given by

$$\mathbf{v} = \mathbf{w}_1 - \mathbf{w}_2 \tag{3.1.4}$$

and the relative kinetic energy in the centre-of-mass frame is

$$\begin{aligned} E_{cm} &= \frac{1}{2}m_1\mathbf{w}_1^2 + \frac{1}{2}m_2\mathbf{w}_2^2 \\ &= \frac{1}{2}\mu\mathbf{v}^2 \end{aligned} \tag{3.1.5}$$

where μ is the reduced mass

$$\mu = \frac{m_1 m_2}{(m_1 + m_2)}. \tag{3.1.6}$$

For hyperthermal collision experiments, where the thermal velocity of the target atom/molecule can normally be neglected compared to the high velocity of the projectile, it is convenient to give an expression for E_{cm} in terms of the laboratory collision energy, E_{Lab}:

$$E_{cm} = \frac{1}{2}\frac{m_1 m_2}{(m_1 + m_2)}(v_1 - v_2)^2$$
$$\frac{1}{2}\frac{m_1 m_2}{(m_1 + m_2)}v_1^2 = \frac{m_2}{(m_1 + m_2)}E_{Lab} \tag{3.1.7}$$

where the subscript 1 refers to the fast projectile and 2 to the slow target.

Equation (3.1.7) implies that the amount of energy available for internal excitation in a hyperthermal atom -fullerene collision will depend very much on whether the fullerene is the projectile or the target. If a fast fullerene ion collides with a slow He target the centre-of-mass collision energy is only 0.55% of the laboratory kinetic energy whereas if a fast He collides with a thermal fullerene target then the centre of mass collision energy is 99.4% of the laboratory kinetic energy.

The conversion between scattering angles and scattering intensities in the lab and centre-of-mass frames of references is most conveniently made with the help of Newton diagrams, Fig. 3.1. In the laboratory system, the centre-of-mass moves with a constant velocity $\mathbf{v}_{cm}$, which is unchanged by the collision. In the centre-of-mass system, the centre-of-mass is at rest and the deflection θ_{cm} is the deflection of the relative velocity vector $\mathbf{v}'$. In elastic collisions, the magnitude of the relative velocity does not change. The final velocity of a particle in the laboratory frame of reference is given by a vector addition of its final relative velocity with respect to the centre-of-mass and the velocity of the centre-of-mass itself. The particle speeds with respect to the centre-of-mass also do not change

$$w_1' = m_2 v' / (m_1 + m_2)$$
$$w_2' = m_1 v' / (m_1 + m_2) \tag{3.1.8}$$

However, the final laboratory velocities are different from their initial values in both magnitude and direction

$$\mathbf{v}_1' = \mathbf{v}_{cm} + \mathbf{w}_1'$$
$$\mathbf{v}_2' = \mathbf{v}_{cm} + \mathbf{w}_2' \tag{3.1.9}$$

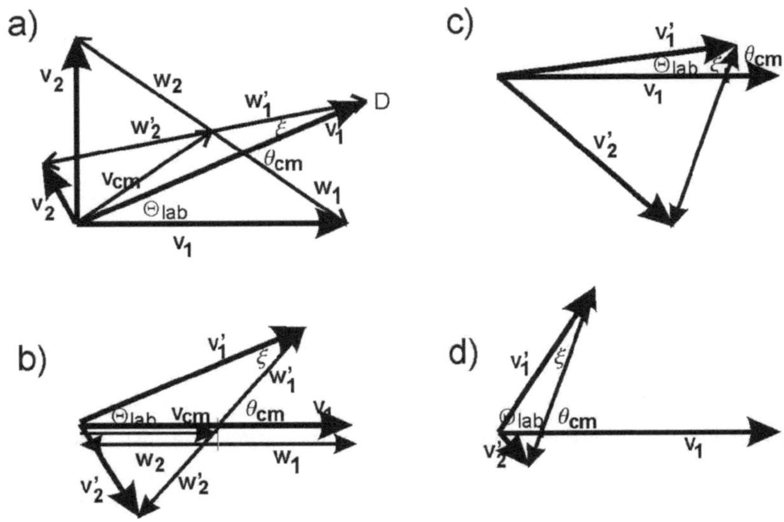

Figure 3.1. Newton diagrams illustrating the conversion between laboratory and centre-of-mass coordinates for collision processes. (a) elastic collision between two particles with initial velocities v_1 and v_2 impinging at 90° and with equal masses. (b) as (a) but with particle 2 initially at rest. (c) Inelastic collision between a fast C_{60} projectile and a stationary Xe atom where half of the centre-of-mass energy is converted to internal excitation. (d) the same centre-of-mass collision energy, energy loss and centre-of-mass projectile scattering angle but this time for a fast Xe projectile colliding with a stationary C_{60}.

Fig. 3.1a gives an example of an elastic collision between two particles intersecting at 90° with initial velocities $\mathbf{v}_1$, $\mathbf{v}_2$ and masses $m_1 = m_2$. Particle 1 is scattered through an angle Θ_{lab} in the laboratory frame of reference. This corresponds to a centre-of-mass scattering angle of θ_{cm}. We now need to convert the centre-of-mass scattering intensity (which is what is normally calculated from model potentials) $I_{cm}(\theta_{cm})$ to the corresponding laboratory intensity $I_{lab}(\Theta_{lab})$. For each θ_{cm}, conservation of intensity gives

$$I_{lab}(\Theta_{lab})d\Omega_{lab} = I_{cm}(\theta_{cm})d\omega \qquad (3.1.10)$$

where $d\Omega_{lab}$ is the solid angle element in the laboratory system corresponding to $d\omega_{cm}$ in the centre-of-mass system. The "Jacobian", or solid angle ratio of the transformation $d\omega_{cm}/d\Omega_{lab}$ can be read-off from Fig. 3.1a to be

$$\frac{d\omega_{cm}}{d\Omega_{lab}} = \frac{\mathbf{v}_1'^2}{\mathbf{w}_1'^2 \left|\cos\xi\right|}. \tag{3.1.11}$$

If more than one centre-of-mass angle corresponds to a given laboratory scattering angle this procedure is repeated for each θ_{cm} and the centre-of-mass intensities are summed to yield the total observable laboratory intensity $I_{lab}(\Theta_{lab})$.

Further examples of Newton diagrams for elastic and inelastic scattering are given in Fig. 3.1.

1.1 Collisional Energy Transfer

The simplest possible situation is a collision between two structureless bodies that interact via a central conservative force. In this case, no energy can be exchanged between the relative motion of the particles and any internal modes. Collisions of this type are called *elastic*. The only outcome of the collision is that the particles deflect one another from their original straight-line trajectories, as discussed above. *Inelastic* collisions, on the other hand, involve the transfer of energy from or to internal modes while in *reactive* collisions chemical bonds are broken or formed. At low energies, collisions between rare gas atoms are elastic but for molecules the collisions may induce changes in internal energy or cause chemical rearrangement. In order to fully describe the excitation mechanisms and collision outcome one needs to have accurate potential energy surfaces and coupling constants, which is unrealistic at present for a complex molecule as large as C_{60}. Instead, much simpler considerations are generally made to model the energy transferred in the collisions.

One extreme simplification which seems to work rather well for low energy atom-fullerene collisions is that the atomic collision partner only "sees" a small part of the fullerene cage and undergoes an elastic collision with this[1]. The kinetic energy transferred to the part of the cage involved in the collision is then subsequently transferred to kinetic energy of the other atoms in the cage corresponding to excitation of many vibrational modes (normally a fast equilibration of energy among the vibrational degrees of freedom is assumed). This is known as a "spectator" model [2]. The idea is illustrated in Fig. 3.2 for an atom-fullerene collision where the fullerene is initially at rest. The atom, A, collides with just a small part of the cage, B -

[1] For convenience in the discussion that follows we will normally regard the atom (or atomic ion) collision partner as being the projectile and the fullerene molecule as the target. Of course, exactly the same arguments hold for the opposite case: fast fullerene ions colliding with stationary atomic targets.

for C_{60} this could reasonably be assumed to be e.g. a carbon atom or a C_2 molecular unit - and the rest of the carbon cage, C, is simply a spectator to the collision. Any change in the velocity of B due to an elastic collision with A is thus a change in the relative velocity with respect to C and hence a change in the vibrational energy of the fullerene. The initial momentum $\mathbf{p}_A$ is partitioned after the elastic collision between A and B so as to conserve energy and momentum. The additional assumption is made that the collision is collinear i.e. we consider only a head-on collision between A and B. The velocity of B after the collision is then

$$\mathbf{v}_B' = 2\mathbf{p}_A / (m_A + m_B) \tag{3.1.12}$$

The vibrational excitation is just given by the change in the relative velocity of B and C

$$\begin{aligned} \Delta E_{vib} &= \frac{1}{2}\mu_{BC}\left(\mathbf{v}_B'\right)^2 \\ &= \frac{1}{2}\frac{m_B m_C}{(m_B + m_C)}\frac{4\mathbf{p}_A^2}{(m_A + m_B)^2} \end{aligned} \tag{3.1.13}$$

The ratio of vibrational energy transfer to initial centre-of-mass collision energy is thus

$$\begin{aligned} \frac{\Delta E_{vib}}{E_{cm}} &= \frac{1}{2}\frac{4\mathbf{p}_A^2 m_B m_C}{(m_A + m_B)^2 (m_B + m_C)}\frac{2m_A(m_A + m_B + m_C)}{\mathbf{p}_A^2 (m_B + m_C)} \\ &= \frac{4m_A m_B m_C (m_A + m_B + m_C)}{(m_A + m_B)^2 (m_B + m_C)^2}. \end{aligned} \tag{3.1.14}$$

This expression is plotted in Fig. 3.3 as a function of m_B for three different atoms colliding with C_{60}: He, Ar and Xe. The energy transfer efficiency is always highest when the unit of the carbon cage taking part in the collision has the same mass as the collision partner.

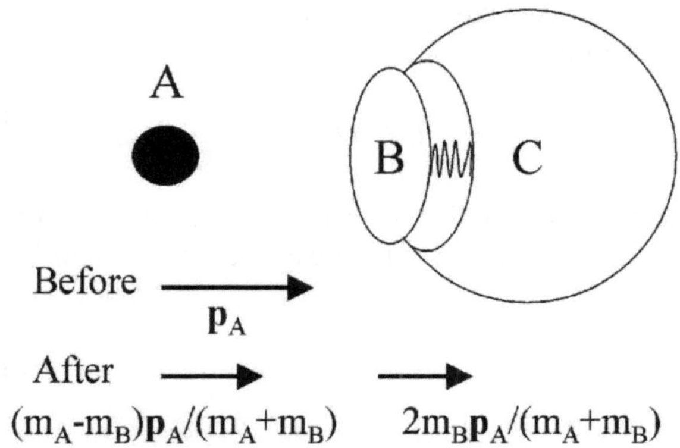

Figure 3.2. Spectator Model

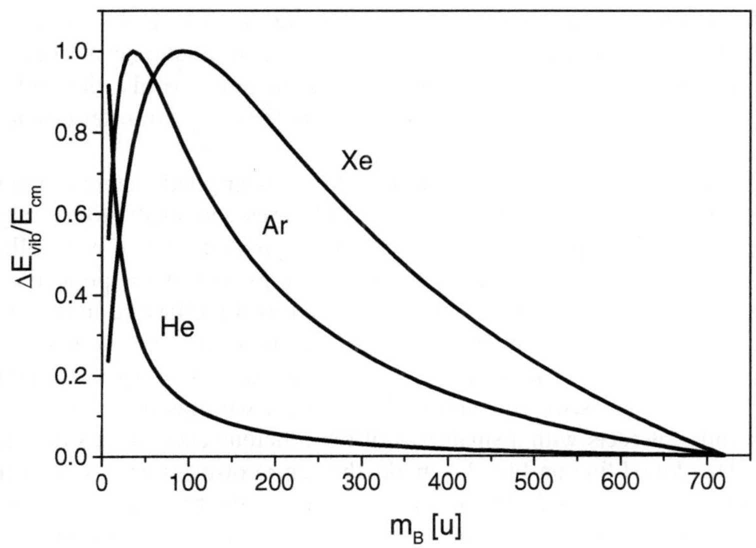

Figure 3.3. Energy loss for C_{60}-rare gas collisions, from the spectator model.

The spectator model regards parts B and C of the molecule to be connected by a very loose spring. So, when the colliding atom A interacts with B, eq. (3.1.10) gives an upper limit for the fractional energy transfer. In physical terms, the limit of a weak oscillator spring is when the duration of the collision, t_c, is short compared to the oscillation period, t_v [2]. We can estimate the duration of the collision as

$$t_c = a / \mathrm{v} \tag{3.1.15}$$

where a is a typical interaction distance and v is the relative velocity during the collision. The spectator model can be expected to be valid for relative collision velocities

$$\mathrm{v} \geq a / t_v = a\nu \tag{3.1.16}$$

where ν is the oscillation frequency.
If we take the average C_{60} vibration frequency, ν = 2.79×10^{13} s^{-1}, and the diameter of the fullerene cage to estimate the interaction distance, a = 10 Å, then we obtain $\mathrm{v} \geq 27900\ ms^{-1}$. For collisions in which energetic C_{60} ions collide with stationary targets this corresponds to a laboratory kinetic energy $E_{Lab} \cong 2900$ eV.

At lower collision velocities the duration of the collision is long compared to the oscillation period. In this case the oscillation can "adjust" to the perturbation caused by the collision or, in other words, the collision partner "sees" the entire molecule. In this limit, the energy transfer efficiency is very low.

Evidence that this simple model provides a rather good picture of the low energy hyperthermal collisions between fullerenes and small atomic targets was obtained in experiments in which a C_{60}^+ projectile beam was collided with a stationary He target [5]. The kinetic energy lost by the projectile ion was measured in the experiments and related to the effective mass on the fullerene cage, m_B, according to eq. (3.1.13). The results are plotted in Fig. 3.4. At low centre-of-mass energies, corresponding to relative velocities $\mathrm{v} < a\nu$, the atom "sees" the entire fullerene cage whereas for velocities the atom only interacts with a small part of the fullerene cage, in this case a C_2 unit. The dotted line on Fig. 3.4 marks the centre-of-mass energy for which the relative velocity is 27900 ms^{-1} and thus marks the boundary between the high and low velocity limits, according to eq. (3.1.16). The agreement with the experimental results is surprisingly good.

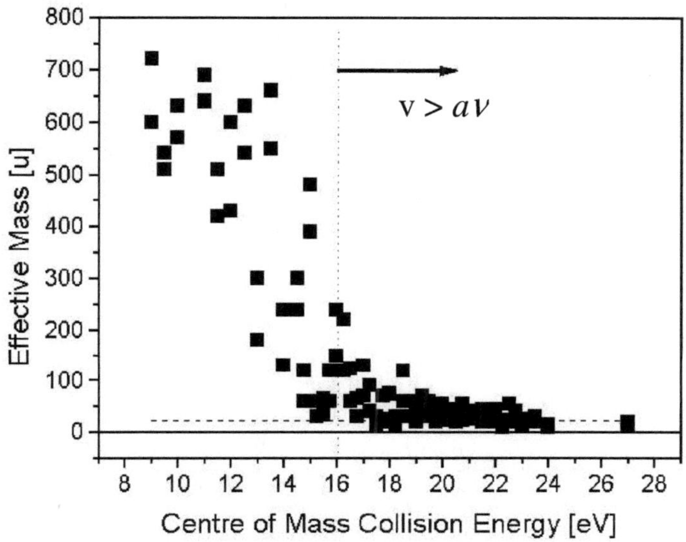

Figure 3.4. Experimentally determined effective mass on the fullerene cage that interacts with a colliding He atom, plotted as a function of centre-of-mass collision energy [5] The dotted line marks the separation between the low velocity regime and the high velocity regime where the spectator model is expected to give a good approximation.

This simple model is fairly good for velocities beyond those given by eq. (3.1.16) but not indefinitely. As the interaction time during the collision becomes shorter and shorter the energy transfer efficiency due to elastic collisions with individual atoms or small molecular units decreases again. For the high energy limit, where the collision duration is comparable to the period of the electron orbits in the target then the energy is not efficiently transferred to the nuclear motion of the target (vibrational excitation) but instead there is predominantly a direct electronic excitation. This is often referred to as the crossover from the "nuclear stopping" regime to that of "electronic stopping". This terminology is taken from the ion implantation community and, to a reasonably good approximation, one can treat fullerene target molecules as a very small double layer of graphite and use the models developed for describing energetic ion implantation of materials to estimate the extent of energy transfer.

The collisions in the "low energy" limit, i.e. where nuclear stopping dominates, are treated in a fashion similar to that discussed above for the simple spectator model. The collisions are treated as binary elastic collisions between the atomic projectile and individual atoms in the target. The main difference is that the collisions are not restricted to collinear, zero impact parameter collisions but the scattering due to the interaction potential of the projectile and target is also taken into consideration and multiple binary collisions can occur during passage of the projectile atom through the target molecule. The nuclear stopping is most simply and generally described by a suitable scaling of energy and cross-section as described in detail by Lindhard et al. [6]. Here, the dimensionless quantities

$$\varepsilon = E\frac{aM_2}{Z_1Z_2e^2\left(M_1+M_2\right)} \tag{3.1.17}$$

and

$$\rho = RNM_2 4\pi a^2 \frac{M_1}{\left(M_1+M_2\right)^2} \tag{3.1.18}$$

are measures of collision energy and range. Z_1, M_1 and Z_2, M_2 are the atomic numbers and mass numbers of the projectile ion and target atom, respectively, E is the projectile ion kinetic energy. R is the range of the interaction (dependent upon the interaction potential), N is the number of atoms per unit volume and $a = 0.8853 a_0 Z^{-1/3}$, a characteristic screening length where $Z^{2/3} = Z_1^{2/3} + Z_2^{2/3}$ and a_0 is the Bohr radius[2]. The nuclear stopping power is defined (in atomic units) as

$$S = \frac{d\varepsilon}{d\rho}\frac{4\pi a Z_1 Z_2 e^2 M_1}{\left(M_1+M_2\right)} \tag{3.1.19}$$

The derivative $d\varepsilon/d\rho$ is a dimensionless measure of the reduced stopping power that allows all nuclear stopping cross sections to be described by one curve. Fig. 3.5 shows the computed reduced stopping power for two common forms of the interaction potential: a Thomas-Fermi type potential (dashed line) and a screened Bohr potential (dotted line). Details of the calculation can be found in Lindhard et al.[6].

The ion atom potential must have the form

[2] Atomic units are used throughout this section unless otherwise stated.

$$V(\mathbf{R}) = \frac{Z_1 Z_2 e^2}{R} u \tag{3.1.20}$$

where the function u must tend to unity when $R \to 0$, and must vanish when $R \to \infty$, because the Coulomb potential is screened at large distances. The screened Bohr potential assumes a power law potential, $V(R) \propto R^{-s}$ or

$$u = \frac{k_s}{s}\left(\frac{a}{R}\right)^{s-1} \tag{3.1.21}$$

with s = 2 and k_2 = 2/(0.8853e). The slightly more realistic simplified Thomas-Fermi potential describes the static interaction between two atoms as $u = u(R/a)$. The classical deflection angle, and thus the elastic scattering cross section, can be shown in this case to depend only on the reduced parameters ε and ρ [6]. The reduced stopping power calculated from the Thomas-Fermi potential, Fig. 3.5, clearly shows a maximum in the nuclear stopping with a rapid fall in energy transfer efficiency due to elastic scattering as the reduced collision energy increases beyond this maximum.

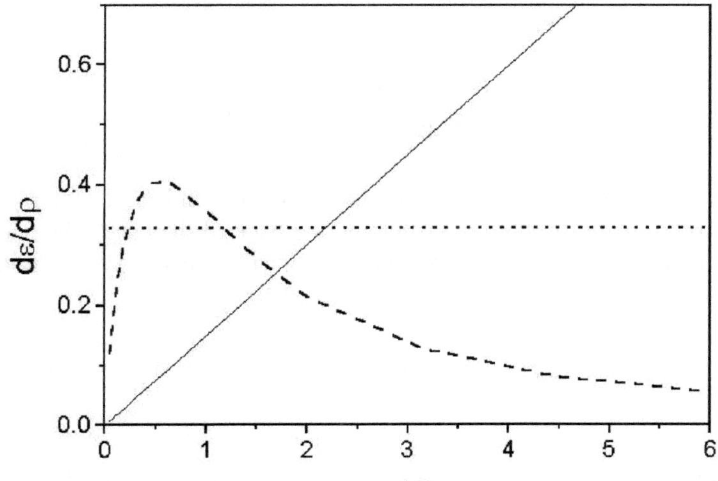

Figure 3.5. Reduced stopping power as a function of the square root of the reduced collision energy (proportional to projectile velocity). Dashed: Thomas-Fermi interaction potential, Dotted: screened Bohr potential, Full: electronic stopping. After Lindhard et al. [6].

The electronic stopping is approximately proportional to the projectile velocity for velocities that are low compared to the orbital velocities of electrons which could be carried by the ion, i.e. according to the Thomas-Fermi estimate [6]

$$\mathrm{v} < \mathrm{v}_1 = Z_1^{2/3}\mathrm{v}_0 \tag{3.1.22}$$

where $\mathrm{v}_0 = e^2/\hbar$. In this range, the electronic stopping power can be approximated by

$$\frac{d\varepsilon}{d\rho} = k\varepsilon^{1/2} \tag{3.1.23}$$

with $k = 0.15$ [7]. This has also been plotted on Fig. 3.5 (full line). Electronic and nuclear stopping are of the same order of magnitude when $\mathrm{v}_1 \approx 0.1\mathrm{v}_0$. or $\varepsilon^{1/2} \approx 2$. For higher projectile velocities the electronic stopping rises sharply to a maximum and then falls off again as $1/\mathrm{v}^2$ [8].

Let us now consider some absolute values for the energy transfer in collisions between projectile ions and C_{60}. The stopping can be converted to the energy transfer to internal modes of the fullerene in eV by making some additional assumptions. We regard only central collisions and regard the C_{60} as a thin carbon foil with a density

$$D = 2\frac{60\cdot 12u}{4\pi r^2} = 0.15\mu g/cm^2 \tag{3.1.24}$$

where r = 3.55 Å is the radius of the fullerene cage and the factor 2 takes account of collisions at both the front and the back of the cage. The energy lost by the projectile ion in eV (assuming interaction occurs with individual C atoms in the cage) is then

$$\Delta E = D\left[\mu g/cm^2\right]S\left[eV/(\mu g/cm^2)\right]. \tag{3.1.25}$$

This is plotted as a function of projectile kinetic energy in Fig. 3.6 for H^+, He^+, Ar^+ and Xe^+ projectile ions colliding with C_{60}, using a simplified Thomas-Fermi potential to describe the nuclear stopping.

This simple picture is not successful at very low collision energies where the Thomas-Fermi approximations are not expected to work very well (e.g. predicting energy transfer that is significantly higher than the collision energy!). In this region, the spectator model is a better approximation, as discussed above. For comparison, we show the predictions of the spectator model, assuming collisions with a single carbon atom on the cage, for the He^+ collisions in Fig. 3.6. However, at energies around or beyond the maximum in the nuclear stopping power the results using the Thomas-Fermi approximation should give reasonable values. They are also in very good agreement with somewhat more sophisticated although similar calculations using the TRIM programme (Transmission and Ranges of Ions in Matter [9]) [10]. Note that the energy transfer efficiency increases strongly with the mass of the projectile ion. Electronic stopping dominates already at relatively low laboratory energies for collisions with protons and helium ions. This also has consequences for the fragmentation patterns observed in experiments as will be discussed in Chap. 5.

Some more sophisticated calculations have been carried out e.g. taking into account the impact parameter dependence [11], using a Monte Carlo method to estimate the nuclear stopping [12] or using an improved treatment of the electronic stopping [13] but the overall picture remains close to that estimated from the Lindhard model, discussed above.

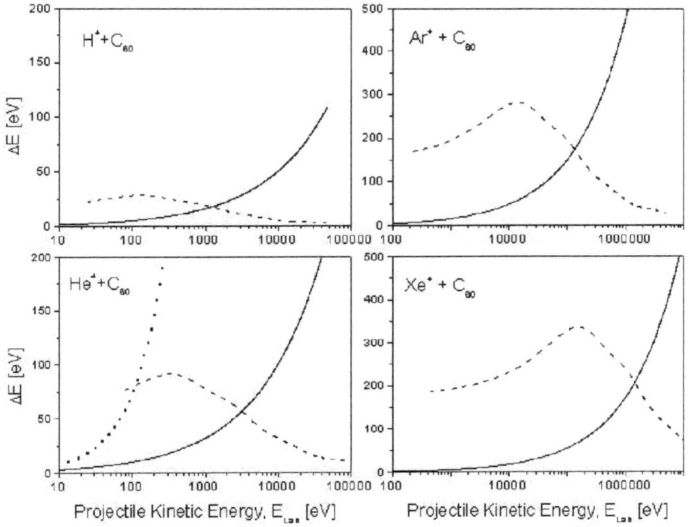

Figure 3.6. Energy transfer to the fullerene cage in collisions with energetic projectile ions, as a function of ion kinetic energy. Dashed line: nuclear stopping calculated using a simplified Thomas-Fermi potential [6], full line: electronic stopping (eq. 3.1.23). The estimate using the spectator model for He+ colliding with an individual C atom on the fullerene cage, (eq. 3.1.14) is shown for comparison (dotted line). Note the different scales on the left and right sides of the figure

1.2 Scattering Dynamics

There have been relatively few studies of the scattering dynamics of fullerenes so far. For hyperthermal collision energies it is possible to make use of simple classical models of the interaction behaviour for heavy particle collisions (i.e. atom-fullerene, ion-fullerene, molecule-fullerene or fullerene-fullerene).

Elastic collisions can be described by the following relations. For a fixed centre-of-mass collision energy E, interaction potential $U(r)$ and impact parameter b, the centre-of-mass deflection angle in an elastic collision is given by the classical deflection function [1, 2]

$$\chi(E,b) = \pi - 2b\int_{r_0}^{\infty} \frac{dr/r^2}{\sqrt{1-U_{eff}(r)/E}} \tag{3.1.26}$$

where U_{eff} is the effective interaction potential given by the sum of the interaction potential and the centrifugal energy and r is the separation of the colliding particles

$$U_{eff}(r) = U(r) + E\left(\frac{b}{r}\right)^2. \tag{3.1.27}$$

The expression for the differential cross section can be obtained from the deflection function and is written as

$$\sigma(|\chi|,E) = \left|\frac{b}{(d\chi/db)\sin\chi}\right|. \tag{3.1.28}$$

The expression for the total elastic scattering cross section is given by the integration of eq.(3.1.28) over all scattering angles

$$\sigma_{total}(E) = \int_0^{\pi} \sigma(\chi,E)2\pi\sin\chi d\chi. \tag{3.1.29}$$

Thus by using a suitable interaction potential *U(r)* it is possible to calculate the cross sections and to compare them with those obtained from the experiments. Alternatively, if the experiments have sufficient resolution one would use the experimentally determined differential cross sections to determine the interaction potential. Unfortunately the experiments are not yet sufficiently advanced to allow us to do this. Also, there are many competing reaction channels in fullerene collisions that make the situation rather complicated [14].

Some estimates of interaction potentials have been made in the literature for fullerene-atom collisions, in particular with reference to Penning ionisation experiments [15] in order to estimate the expected magnitude of the cross sections and probability of energy transfer to compare with the experimental results.

Presently the most reliable interaction potential was calculated by Girifalco for fullerene-fullerene collisions using a Lennard-Jones type potential [16, 17]. In this potential, the C-C interactions within the C_{60} molecule are averaged over the surface of a sphere. This reduces each C_{60} molecule to a pseudo-atom with just 3 centre-of-mass coordinates and introduces rotational isotropy. The potential has the form

$$U(r) = -\alpha\left[\frac{1}{s(s-1)^3}+\frac{1}{s(s+1)^3}+\frac{2}{s^4}\right]+\beta\left[\frac{1}{s(s-1)^9}+\frac{1}{s(s+1)^9}+\frac{2}{s^{10}}\right] \tag{3.1.30}$$

where the parameters are $\alpha = 0.04678$ eV, $\beta = 8.486 \times 10^{-5}$ eV and $s = r/7.1$ Å. The simplicity of the Girifalco potential is one of its advantages. Although it was shown that the Girifalco potential underestimates the stability of the liquid phase of C_{60} [18, 19] it is relevant for the solid phase. It is also consistent with elastic scattering measurements for $C_{60}^{+} + C_{60}$ gas phase experiments [14] when an additional polarisability term is added

$$U_{pol}(r) = -\frac{\eta e^2}{2r^4} \tag{3.1.31}$$

with the polarisability of C_{60}, $\eta = 80 \pm 4$ Å^3 [20, 21].

The Girifalco interaction potential (eq. 3.1.30) is plotted in Fig. 3.7. The potential rises very rapidly as the fullerenes approach each other and is quite close to a hard sphere potential. Due to the rather unusual properties of the intermolecular potential it is not possible to use the conventional unique small angle scattering relationship between $E\chi$ and b [22] even for relatively high collision energies of a few hundred eV. Instead, the deflection function has to be calculated for each collision energy [14]. At the relatively high collision energies that have been studied so far (180-1500 eV) there is very little influence of the shallow potential well and the interaction probed in the collisions is purely repulsive. Lower energy collisions, although difficult to carry out experimentally, should be feasible and would provide a good and needed test of this simple but useful interaction potential.

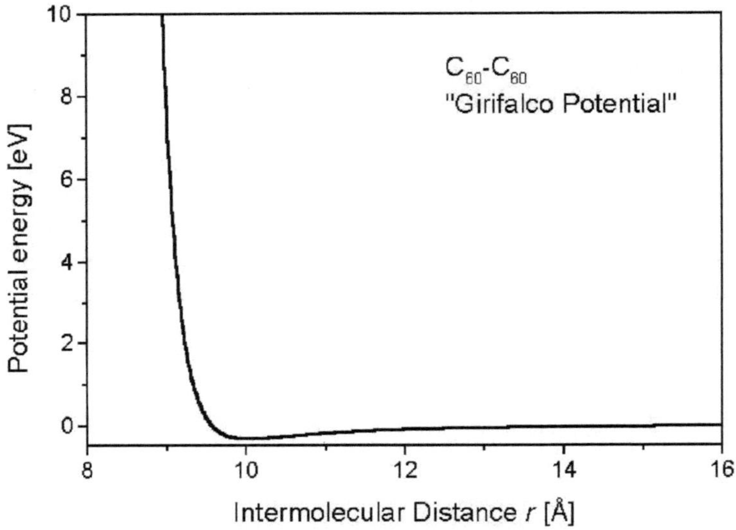

Figure 3.7. "Girifalco" potential [16, 17] for the C_{60}-C_{60} interaction

2. FRAGMENTATION: STATISTICAL MODELS

Generally the fragmentation behaviour of collisionally excited fullerenes can be reasonably well understood in terms of simple statistical models. In these models it is invariably assumed that the excitation energy is rapidly statistically distributed among all internal degrees of freedom of the molecule. This seems to be a good assumption for relatively low collision energies where the energy transfer is predominantly via vibrational excitation (nuclear stopping) and where there is sufficient time for energy redistribution to occur (> ps) before significant fragmentation takes place. However, the assumptions are more questionable for high energy collisions dominated by electronic stopping or where there is simply not sufficient time for equilibration to occur before the molecule disintegrates. Some reports of non-statistical fragmentation have been made and will be discussed in chapter 5, however, the evidence for non-statistical behaviour is still rather limited and in some cases the results can also be consistent with a statistical model.

Here we present the simplest and most commonly used statistical models applied to the fragmentation of fullerenes.

2.1 Arrhenius Fragmentation Rates and the Finite Heat Bath Correction

A highly excited fullerene can release its energy in different ways; it can fragment, it can emit electrons or it can emit photons. One of the most interesting aspects of excited neutral fullerenes is that all three decay channels compete with each other for excitation energies and time scales typical of many experiments [23]. This is unusual for large molecules and is a consequence of the high dissociation energy combined with a relatively low ionisation potential as well as the ability of the molecule to "store" large amounts of energy for relatively long times due to the large number of vibrational degrees of freedom. For positively charged fullerenes the situation is simpler since statistical electron emission ("thermionic emission") does not compete significantly with fragmentation and radiative cooling due to the high value of the second ionisation potential (11.4 eV). The importance of competing statistical decay channels has not always been recognised and this has contributed to the difficulty of obtaining a reliable value for the dissociation energy for C_{60} i.e. the energy required for the reaction $C_{60} \rightarrow C_{58} + C_2$. Values for this energy reported in the literature vary over a large range [24] and there is still a certain amount of controversy and uncertainty concerning the correct value. A further problem with determining this value is that it cannot be measured directly in an experiment. Most experimental determinations involve invoking a statistical model and making some assumptions about the mechanism such as the looseness of the transition state [25, 26]. Until recently the experimental values were significantly below estimates obtained from high-level density functional calculations [27], however, the experimental values are creeping upwards and the values being reported are now compatible with the most reliable theoretical estimates [24, 25, 28-31]. In the following, where necessary, we will use values that are presently considered to be the most appropriate to describe the metastable fragmentation (or C_2 evaporation) of fullerenes, however, it should be noted that this may change as our knowledge of the complex mechanisms involved increases. The present status of the dissociation energies of the ions is summarised in Fig. 3.8.

Although the absolute values are still under investigation the relative values of the dissociation energies should be reliable.

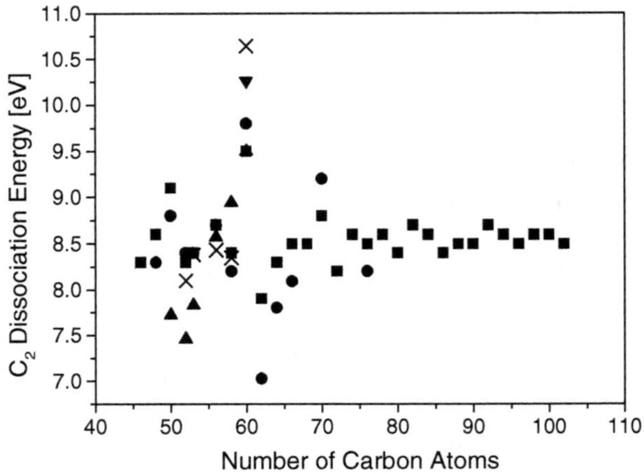

Figure 3.8. Dissociation energies for fullerene cations. Experimental: Circles from [30]. Up-triangles from [29]. Down triangle from [32]. Squares are relative values from [33] that have been normalised to the n=54 value of Tomita et al. [30]. Theory: Crosses from [31].

The dissociation energies of the neutral fullerenes are related to the fullerene cation dissociation energies according to the scheme shown in Fig. 3.9. The dissociation energy for C_2 loss from C_{60} is thus 0.54 eV higher than that from singly ionised C_{60}. The ionisation potentials of the fullerenes are tabulated in Table 3.1.

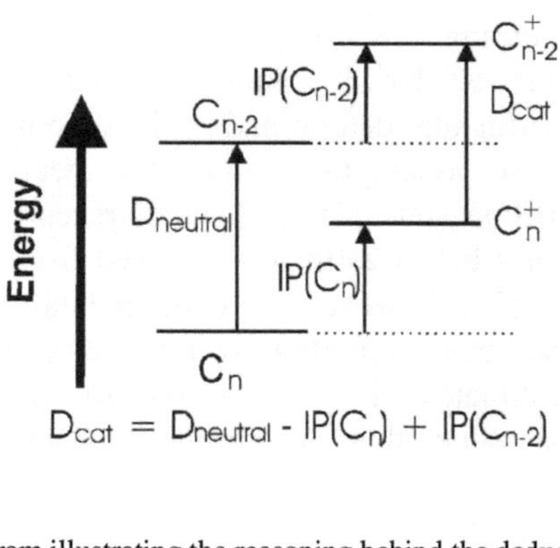

Figure 3.9. Diagram illustrating the reasoning behind the deduction of the cation C_2 dissociation energy, D_{cat}, from that of the corresponding neutral fullerene value, $D_{neutral}$.

Table 3-1. Ionisation Potentials of the Fullerenes

Fullerene	IP [eV]	Reference	Fullerene	IP [eV]	Reference
C_{48}	7.17±0.1	[34]	C_{60}^{+}	11.4±0.04	[35]
C_{50}	7.61±0.11	[34]	C_{60}^{2+}	16.6±1.0	[36]
C_{52}	7.07±0.1	[34]		15.6±0.5	[37]
C_{58}	7.07±0.1	[34]		17.0±0.75	[38]
C_{60}	7.59±0.02	[39]	C_{60}^{3+}	20±1.0	[40]
C_{62}	7.07±0.1	[34]	C_{60}^{4+}	22.3	[41]
C_{66}	7.27±0.14	[34]	C_{70}^{+}	11.4±0.06	[35]
C_{68}	7.27±0.14	[34]	C_{70}^{2+}	16.5±2.0	[42]
C_{70}	7.48±0.05	[39]	C_{70}^{3+}	19.2±5.0	[42]
C_{72}	6.97±0.1	[34]	C_{70}^{4+}	29.2±8.3	[42]
C_{74}	6.84±0.1	[34]			
C_{76}	7.1±0.1	[39]			
C_{78}	7.05±0.1	[39]			
C_{80}	6.84±0.1	[34]			
C_{84}	7.15±0.1	[39]			
$C_{100-200}$	6.2-6.45	[34]			

The simplest description of the dissociation rates is given by the Arrhenius-like equation

$$k_{diss} = A_{diss} \exp\left(\frac{-D}{k_B T}\right) \tag{3.2.1}$$

where T is the internal temperature of the fullerene and D is the dissociation energy. A similar expression is used to describe the delayed "thermionic" emission of electrons with D given by the ionisation potential and A_{Diss} replaced by the appropriate pre-exponential factor for ionisation [43].

The appropriate parameters to use in order to describe the C_2 emission from C_{60} neutrals, ions and other fullerene species are still a matter of investigation, as mentioned above. The most recent experimental determination of the dissociation energy for C_2 loss from neutral C_{60} places it at 10.8±0.3 eV [32], in good agreement with theoretical values (ca. 11-11.5 eV [27, 31]). This corresponds to a dissociation energy for the singly charged cation of ca. 10.3 eV. The present status of our knowledge concerning the dissociation energies of the singly charged fullerene cations is plotted in Fig. 3.8. Values for the dissociation energies of doubly charged cations can be found in [31] and references given there. C_{60} has an exceptionally high dissociation energy, compared to the other fullerenes but, even more unusual is the extremely high value of the pre-exponential factor A_{Diss} that is needed for C_{60} and the other fullerenes in order to obtain dissociation rates that are compatible with experiment [25, 30]. The most recent evaluation and modelling of experimental data provides a pre-exponential factor of $3.4\text{x}10^{21}$ s^{-1} for C_2 loss from C_{60} with somewhat lower values for the other fullerenes [32]. Theoretical estimates can obtain values within the range 10^{20}-10^{21} [25, 44] but for many researchers it is intuitively difficult to accept such high values and the fragmentation mechanism is still under debate. A simplified expression for the temperature dependence of the C_{60} dissociation pre-exponential factor that is often used is [25]

$$A_{Diss} = 3.0\times10^{12}T^{3}\left(1-\exp\left(-1441/T\right)\right)^{5} s^{-1} \qquad (3.2.2)$$

The temperature commonly used in eq.(3.2.1) and (3.2.2) is not simply the vibrational temperature of the molecule but is a "heat bath" temperature. This is the temperature that a heat bath must be set to so that the canonical rate constant *k(T)* is equal to the micro-canonical rate constant *k(E)*[3]. The finite heat bath correction [45] relates the heat bath temperature to the vibrational temperature

$$T_{vib}(E) = T(E)\left[1+\frac{1}{2}\frac{D}{TC}+\frac{1}{12}\left(\frac{D}{TC}\right)^{2}+....\right] \qquad (3.2.3)$$

where T_{vib} is the vibrational temperature of the fullerene, C is one less than the heat capacity of the emitting species at temperature T (in units of k_B) and D is the activation energy of the reaction under consideration.

It is generally accepted that the most significant fragmentation channel of an excited fullerene at modest excitation energies is the successive loss of C_2 units (with the exception of fission reactions of highly charged fullerenes as discussed in Chapter 6). This is also the channel that is observed in experiments that study the metastable decay of fullerenes on the microsecond time scale. However, for high levels of excitation it is possible that the fullerene cage rapidly loses its structure (on the sub-ps time scale) [46] and it is then energetically more favourable for it to break up into ring and chain fragments. The cross-over from predominantly C_2 evaporation to rapid break-up into two or more larger fragments shows some signatures of a

[3] A *micro-canonical* ensemble is an ensemble of sub-systems where each sub-system has a fixed energy. A *canonical* ensemble is an ensemble of sub-systems where energy can flow between the sub-system and a heat reservoir so that the temperature is fixed but the energy of each sub-system can fluctuate.

phase transition [46, 47]. This is also seen in maximum entropy calculations, discussed in the next section.

2.2 Maximum Entropy Calculations

The maximum entropy model calculates the most probable distribution of a collection of atoms using the heats of formation ΔH_j, the ionisation potentials, the number of isomers g_j and the mean vibrational frequencies of all possible species that can be formed from the collection of atoms. It was developed to describe the fragmentation patterns of organic molecules [48] and has also been applied to atomic cluster fragmentation [49] and to nuclear collisions. The most probable distribution is given by finding the maximum entropy of the system. The only constraints present are the conservation of energy, mass and charge. No assumptions whatsoever are made concerning the fragmentation mechanism, only that the energy is statistically distributed among all degrees of freedom. The model has been applied to the fragmentation of C_{60}, C_{60}^+ [50, 51], La@C_{82}^+ [51] and C_{120}^+ [52]. The simple model gives remarkably good agreement with experiment for fullerene fragmentation - much better than for organic molecules. This may be partly a consequence of the relative simplicity of the fullerenes, consisting of only one atomic element, but is also a strong sign of the highly statistical nature of the fullerene fragmentation. Starting with a given temperature of the fragments, the model iterates through the translational, vibrational and rotational partition functions to find the maximum total entropy. It can be shown [49] that the abundance X_j of species *j* is given by:

$$X_j = g_j Q_j V \exp\left(-\beta \Delta H_j^{\#} - \gamma_n n_j - \gamma_c c_j\right) \qquad (3.2.4)$$

where g_j is the number of isomers of species *j*, Q_j is the partition function per unit volume, *V* is the interaction volume, n_j and c_j are the number of atoms and the charge of species *j*, respectively. $\Delta H_j^{\#}$ is the heat of formation relative to that of the parent species $\Delta H_j^{\#} = \Delta H_j - \Delta H_{parent}$ [4].The Lagrange multipliers γ_n and γ_c and the initial mean internal energy of the parent molecule $\langle E \rangle$ have to be determined in order to find the maximum entropy starting with a given fragment temperature. The mean energy is given by

$$\langle E \rangle = \sum_j X_j \langle \varepsilon_j \rangle \qquad (3.2.5)$$

[4] Note that this was not done correctly in the original application of this model to C_{60} fragmentation [50] but was corrected in later work [51, 52].

The mean energy of species *j* is

$$\langle \varepsilon_j \rangle = \sum_i \varepsilon_{ij} x_{ij} \tag{3.2.6}$$

where ε_{ij} is the energy of quantum state *i* calculated from a common zero of all isomers *j* adding the vibrational, rotational and translational energies to the heats of formation. Note that the heats of formation ΔH_j and ΔH_{parent} are negative. $\Delta H_j^{\#}$ will therefore be a positive number.

$$\varepsilon_{ij} = \varepsilon_i^{vib} + \varepsilon_i^{rot} + \varepsilon_i^{trans} + \Delta H_j^{\#} \tag{3.2.7}$$

and the fraction of molecules of species *j* in the quantum state *i* is

$$x_{ij} = Q_j^{-1} \exp\left(-\beta\left(\varepsilon_{ij} - \Delta H_j^{\#}\right)\right) \tag{3.2.8}$$

The partition function can be calculated from standard partition functions for the vibrational, rotational and translational degrees of freedom, given e.g. in the JANAF tables [53]. The conservation of mass and charge provides the constraints

$$N = \sum_j n_j X_j \tag{3.2.9}$$

where $N = 60$ for C_{60} and

$$C = \sum_j c_j X_j \tag{3.2.10}$$

with C=0 for a neutral parent molecules and C=1 for a singly charged cation. The interaction volume V (eq.3.2.4) is a parameter that influences the relationship between the initial internal excitation energy of the parent molecule and the resulting fragment temperature but does not critically affect the qualitative excitation energy dependence of the fragment distributions. For all applications to fullerene fragmentation, V was set to a value equal to approximately three times the volume of the parent molecule.

The heats of formation, ionisation potentials, vibrational frequencies and number of isomers are rather well known for the carbon species from C to C_{60} and, where unknown, it is possible to extrapolate from known values to obtain a reasonable "guestimate". For the parameters used in the calculations see [50] and references given therein. Extrapolations were also made to obtain the input parameters for calculations involving C_{120}^+ [52] and La@C_{82}^+ [51].

The abundance of a particular species predicted by the model depends, to a large extent, on the population density in the available translational and vibrational states. To create smaller fragments, the difference in binding energies between the parent fullerene and the particular fragment has to be overcome. At low energies not enough energy is left for higher lying, more degenerate states to be populated. The larger fullerenes with a lower barrier to overcome are therefore favoured at lower excitation energies. With increasing excitation energy, entropy starts to play a role. The smaller fragments become dominant as they are favoured due to their high number of translational states. The model distributions are therefore obtained as a consequence of competition between these two tendencies. Figure 3.10 shows the results of maximum entropy calculations of the fragmentation of C_{60}^+. The upper part shows the calculated fragment distribution for an initial parent excitation energy $\langle E \rangle$ = 110 eV. The typical fullerene bimodal fragmentation pattern is clearly reproduced. The relative intensities of the various fragments are critically dependent on the heats of formation and ionisation potentials used in the calculation [51]. No attempt has been made so far to optimise this to obtain perfect agreement with experiment (a change of 0.1 eV in binding energy can make a significant difference to the relative intensity of a given species). Instead the literature values and simple extrapolations from these have been used. The main features observed in low energy collision experiments and ns laser excitation work are reproduced. The fragmentation pattern moves from that of sequential C_2 evaporation from fullerene ions at low excitation energies to a bimodal distribution containing fullerenes ions as well as small fragment ions with predominantly ring and chain geometries as the excitation energy increases beyond ca. 70 eV. The magic numbers observed in the experiments for e.g. n = 7, 11, 15 etc. are reproduced. For very high excitation energy only small fragment ions, predominantly C^+ and C_3^+ remain. The lower part of the figure plots the relationship between the mean initial excitation energy $\langle E \rangle$ and the resultant fragment temperature. Here one can clearly see the signature of a phase transition that occurs at a temperature of slightly less than 4000 K. Evidence for a phase transition at such elevated temperatures is also found in molecular dynamics simulations [46] (section 3.3). The excitation energy at which the phase transition onsets corresponds to the threshold energy for production of the small ring and chain fragment ions C_n^+ with $n < 30$.

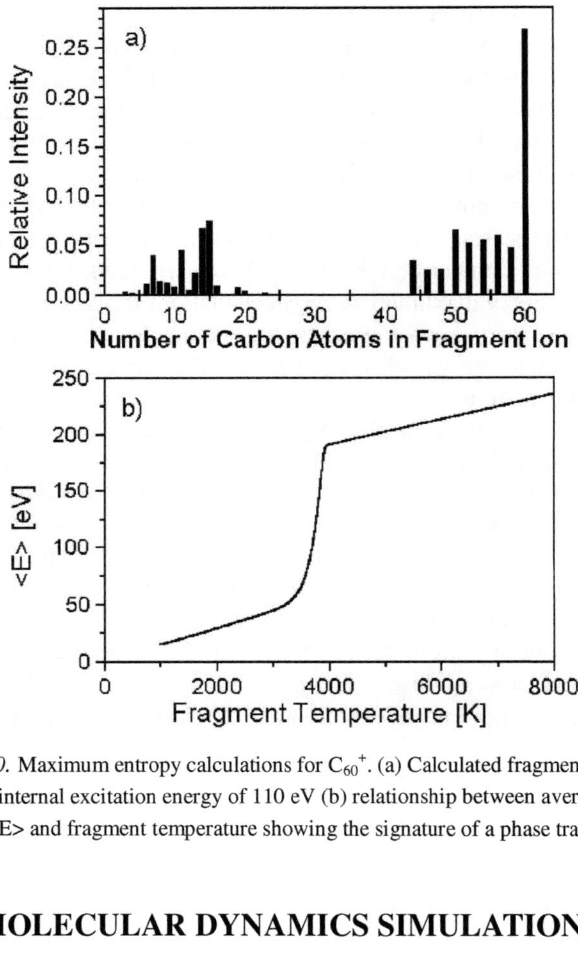

Figure 3.10. Maximum entropy calculations for C_{60}^+. (a) Calculated fragment distribution for an initial internal excitation energy of 110 eV (b) relationship between average internal energy <E> and fragment temperature showing the signature of a phase transition.

3. MOLECULAR DYNAMICS SIMULATIONS

Molecular dynamics (MD) simulations are a very powerful tool to help visualise and intuitively understand the dynamics of collisions and half-collisions. Many studies involving fullerenes have been carried out. It is, however, important to remember that the results of such calculations can be critically dependent on the parameters used in the particular study and one should always be careful not to over-interpret the data however tempting it may sometimes be. It is also the case that MD simulations are generally only valid as long as there is no influence of electronically excited states. It is only very recently that methods are being developed that are able to treat electronic excitation in dynamical studies of fullerenes [54, 55]. Generally, MD simulations involve the calculation of trajectories of an ensemble of particles in space-time. The particles move due to the forces acting on them according to Newton's second law. The equation of motion of a particle of mass m moving in the presence of *(N-1)* interacting particles is thus:

$$\mathbf{F}_i = m\frac{d^2\mathbf{r}_i}{dt^2} = \sum_{\substack{j=1 \\ j\neq i}}^{N} \mathbf{F}(r_{ij}) \tag{3.3.1}$$

where r_{ij} is the distance of the particle i to another particle j and $\mathbf{F}(r_{ij})$ is the force on the particle i exerted by particle j. If one knows the interaction potential at any given time for any given set of coordinates then one can calculate the trajectory of the particle using eq.(3.3.1) and the relation $\mathbf{F}(r) = -\nabla\mathbf{V}(r)$. To realise this on a computer, one makes use of time-discretisation techniques using finite difference methods [56]. One of the most common methods is to make use of the Verlet algorithm [57]:

$$\mathbf{r}_i(t+\delta t) = 2\mathbf{r}_i(t) - \mathbf{r}_i(t-\delta t) + \frac{h^2}{m}\mathbf{F}_i(t) \tag{3.3.2}$$

where δt is a suitable time step, normally on the order of 10^{-15} to 10^{-16} s. The velocity of the atoms can be obtained from

$$\mathbf{v}_i(t) = \mathbf{v}_i(t-\delta t) + \frac{\delta t}{m}\mathbf{F}_i(t) \tag{3.3.3}$$

The major problem of molecular dynamics simulations is to define the correct interaction potential. It is often a very non-trivial task and most of the computation time is allocated to the calculation of the potential. There are different ways of doing this. In the following sections the main methods applied to fullerene collisions will be outlined. General introductions to MD simulation techniques can be found in e.g. [58].

3.1 Classical MD Simulations

Due to the large number of atoms involved in fullerene collisions and the need to calculate many trajectories in order to obtain a statistically relevant picture of the collision dynamics, classical MD simulations have been extensively used. Classical MD simulations are fast, computationally efficient and based on the construction of parameterised, empirical potentials. Carbon has the ability to form many kinds of structures and this must be accounted for in the construction of the empirical potentials, especially if fragmentation processes are to be investigated. The treatment of the covalent character of the carbon-carbon bonds requires the incorporation of three-body forces. One of the most common and successful parameterisations for carbon is due to Tersoff [59]. This works very well for bulk diamond-like structures but is less suitable for treating graphite and is not able to correctly reproduce the structure of the small carbon molecules. This deficiency was overcome by Brenner who generalised the Tersoff potential to be able to treat hydrocarbons and small carbon molecules at the expense of adding more parameters [60] (with corrections to the original paper given in [61]).

The Brenner parameterisation has been found to be very suitable for modelling fullerenes [62]. The carbon-carbon interaction potential has the form

$$V^{C-C} = \frac{1}{2}\sum_{i=1}^{N}\sum_{\substack{j=1\\ j\neq i}}^{N}\left[V_R\left(r_{ij}\right) - B_{ij}V_A\left(r_{ij}\right)\right] \tag{3.3.4}$$

where $V_R(r_{ij})$ and $V_A(r_{ij})$ are the repulsive and attractive parts of the potential, respectively. The parameters for the carbon-.carbon interaction can be written as

$$V_{R,A}\left(r_{ij}\right) = f_C\left(r_{ij}\right)U_{R,A}\exp\left(-\lambda_{R,A}r_{ij}\right),$$

$$B_{ij} = 1 + \sum_{\substack{k=1\\ k\neq i,j}}^{N} f_C\left(r_{ik}\right)G\left(\theta_{ijk}\right),$$

$$G(\theta) = b_0\left[1 + \frac{c_0^{\ 2}}{d_0^{\ 2}} - \frac{c_0^{\ 2}}{d_0^{\ 2} + \left(1+\cos(\theta)\right)^2}\right], \tag{3.3.5}$$

where

$$f_C(r)=\begin{cases}1, & r<R_C-D_C\\ \frac{1}{2}\left[1+\cos\left(\frac{\pi(r-R_C)}{2D_C}\right)\right], & \mathrm{R_C}-D_C\le R_C+D_C\\ 0, & r>R_C+D_C\end{cases}$$

(3.3.6)

B_{ij} is the bond-order function of three-body bond-angle dependent interaction terms, θ is the angle between two bonds, $G(\theta_{ijk})$ is a function of the angle between bonds *i-j* and *i-k*, $f_C(r_{ik})$ restricts the pair potential to nearest neighbours, R_C is the equilibrium distance, D_C is the well depth and $\lambda_{R,A}$ are Morse parameters. The function $G(\theta)$ is parameterised with respect to the binding energies of the C_2 molecule [60]. N is the number of carbon atoms in the system. The values of the various parameters found to be most appropriate for fullerene simulations are given in Table 3-2.

Table 3-2. Brenner parameters used for classical MD simulations involving fullerenes. Taken from [60]. All values in atomic units.

U_R	19.0496
U_A	12.0544
λ_R	1.27498
λ_A	0.98835
b_0	$1.1304\text{x}10^{-2}$
c_0	19
d_0	2.5
R_C	3.496
D_C	0.283459

The above parameterisation and other similar approaches (e.g. [46]) have been used to simulate "half-collisions" in which the dynamical behaviour of a highly excited fullerene is studied as a function of time. They have also been used to study the behaviour of fullerene-fullerene collisions [63] (see Chapter 9).

In many cases one wishes to simulate the dynamical behaviour in collisions between a fullerene molecule and an atom. This is important for understanding the collision energy dependence of fragmentation or for understanding the mechanisms by which atoms are captured in gas phase collisions to produce endohedral fullerenes (see Chapter 8). In this case the interaction between the foreign atom and the carbon atoms is most simply described by a screened Coulomb potential [62, 64]

$$V^{A-C}=\sum_{j=1}^{N}\frac{Z_1Z_2e^2}{r_{Aj}}\chi\left(r_{Aj}/a\right) \qquad (3.3.7)$$

where Z_1 and Z_2 are the charge numbers of the atoms, N is the number of carbon atoms in the fullerene, χ is the Molière screening function [65] and a is an appropriate screening radius. An example of an MD simulation showing capture of a Ne atom by C_{60} at a collision energy of 80 eV is shown in Fig. 3.11. These calculations can provide very useful qualitative information concerning the dynamics of fullerene collisions but are likely to be less reliable when describing the details of the fragmentation behaviour [66]. It should also be remembered that such MD simulations can only reliably and realistically calculate the time dependence on a scale of ps. Experiments, on the other hand, normally detect the products of collisions on a time scale of at least many microseconds after the collision event. Some attempts have been made to combine MD studies with statistical models to reproduce experimental fragmentation cross sections [67].

3.2 MD Simulations using Density Functional Theory

A more reliable but considerably more time consuming treatment uses *ab initio* MD concepts based on density functional theory (DFT). DFT was created in 1964 with the work of Hohenberg and Kohn [68] who proved that the electronic ground state energy and all related properties can be defined from the knowledge of the electronic ground state density alone. Slightly later, Kohn and Sham showed [69] that the exact electronic ground state energy E_0 for a system of N electrons and M nuclei with the ground state electron density ρ can be written as

$$\begin{aligned} E_0 = &-\frac{1}{2}\sum_{i=1}^{N}\langle\psi_i(\mathbf{r})|\nabla^2|\psi_i(\mathbf{r})\rangle - \sum_{A=1}^{M}\int\frac{Z_A\rho(\mathbf{r})}{|\mathbf{r}-\mathbf{r}_A|}d\mathbf{r} \\ &+\frac{1}{2}\int\int\frac{\rho(\mathbf{r})\rho(\mathbf{r}')}{|\mathbf{r}-\mathbf{r}'|}d\mathbf{r}d\mathbf{r}' + E_{exc}\left[\rho(\mathbf{r})\right] \end{aligned} \tag{3.3.8}$$

where the first term is the kinetic energy of the electrons, the second term is the electron-ion interaction energy and the third term is the electron-electron interaction energy. E_{exc} is the exchange correlation energy that depends on

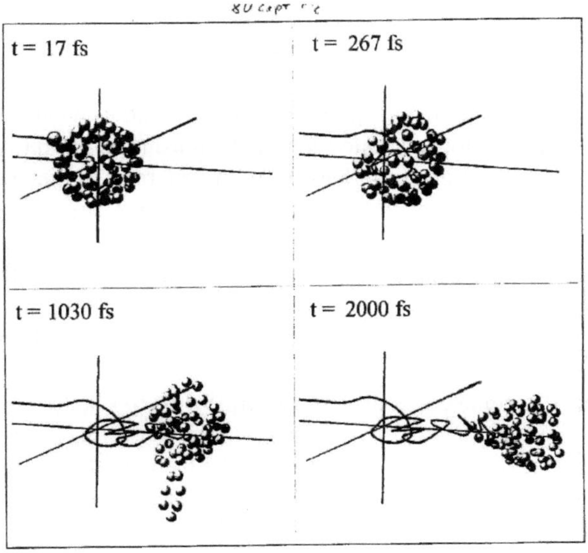

Figure 3.11. Classical MD simulations of the collision of a Ne atom with C_{60} at a collision energy of 80 eV. The Ne atom penetrates the cage and is captured to form a highly excited endohedral molecule. From [70].

the density ρ. The functions $\psi_i(\mathbf{r})$ are the so-called "Kohn-Sham" orbitals. They have no physical meaning but are used to find the ground state density ρ which is given by

$$\rho(\mathbf{r}) = \sum_{i=1}^{N} |\psi_i(\mathbf{r})|^2 . \tag{3.3.9}$$

The Kohn-Sham orbitals are obtained by solving the one-electron equations

$$H_{KS}\psi_i(\mathbf{r}) = \varepsilon_{i,KS}\psi_i(\mathbf{r}) \tag{3.3.10}$$

where H_{KS} is the one-electron Hamiltonian

$$H_{KS} = -\frac{1}{2}\nabla^2 - \sum_{A=1}^{M} \frac{Z_A}{|\mathbf{r}-\mathbf{r}_A|} + \int \frac{\rho(\mathbf{r}')}{|\mathbf{r}-\mathbf{r}'|} d\mathbf{r}' + V_{exc}(\mathbf{r}) \tag{3.3.11}$$

and V_{exc} is the exchange-correlation potential. According to [69], this potential is a functional derivative of E_{exc} with respect to ρ

$$V_{exc} = \frac{\delta E_{exc}[\rho]}{\delta \rho}. \qquad (3.3.12)$$

In the ideal case the exchange-correlation potential handles all the electronic contributions not accounted for by the third term in eq.(3.3.8) and is only a functional of the electron density, ρ. In general, DF methods differ from each other in the form of the exchange correlation energy $E_{exc}[\rho]$. The solution for E_0 is found interatively:

- One starts with an initial guess for ρ and then calculates the initial V_{exc} from eq.(3.3.12)
- The Kohn-Sham orbitals, ψ_i, are numerically calculated using eqs.(3.3.10 and 3.3.11).
- The thus determined ψ_i's are used in eq.(3.3.9) to obtain a new value of ρ,
- One starts again from the beginning of the cycle with the new ρ until the calculation converges. The ground state energy, E_o, is then obtained from the converged ρ using eq.(3.3.8).

The main error in the calculation results from the approximation used for E_{exc}. The local density approximation (LDA) is widely used, whereby $E_{exc}\left[\rho(\mathbf{r})\right]$ depends only on the electron density

$$E_{exc}\left[\rho(\mathbf{r})\right] = E_{exc}\left(\rho(\mathbf{r})\right). \qquad (3.3.13)$$

Calculations in which the electronic states were treated within the LDA with a Fermi-Dirac distribution of the excited levels were carried out for $Li^+ + C_{60}^-$ collisions [71]. These calculations showed the possibility of the direct insertion of the Li ion to form an endohedral fullerene (see Chap. 8). However, there have been very few truly *ab initio* molecular dynamics simulation carried out for fullerene collisions. The high demands on computing capacity that are needed for simulating such large systems have stimulated the development of semi-empirical techniques which can have almost the accuracy of self-consistent LDA techniques but are much less demanding on computer time.

Tight-binding schemes have generally been very successful and can be seen as simplified ab initio methods which still use quantum mechanics to determine the electronic properties of the system. In these calculations the elements of the Hamiltonian matrix are fitted to reproduce an arbitrary set of input data. Normally, only two-centre contributions to the Hamiltonian matrix are considered. However, the fitting can be a rather complicated and difficult procedure. Zhang et al. have applied such calculations to simulate fullerene collisions [72].

Another semi-empirical approach that avoids the problem of fitting the matrix elements has been applied extensively to the study of fullerene collisions [73]. This is based on the work of Seifert, Eschrig and co-workers [74, 75] and applies the formalism of optimised linear combination of atomic orbitals (LCAO). In this approximation, the Kohn-Sham orbitals are expanded in terms of atom-centred localised basis functions $\phi_\mu(\mathbf{r}-\mathbf{R}_J)$ (centred at nucleus *J*).

$$\psi_i(\mathbf{r}) = \sum_\mu c_{\mu i}\phi_\mu(\mathbf{r}-\mathbf{R}_J) \tag{3.3.14}$$

The Kohn-Sham equations (3.3.10) can be transformed into a set of algebraic secular equations

$$\sum_\mu c_{\mu i}\left(h_{\mu\nu} - \varepsilon_i S_{\mu\nu}\right) = 0, \forall \mu, i \tag{3.3.15}$$

where the matrix elements are defined as

$$\begin{aligned} h_{\mu\nu} &= \left\langle \phi_\mu \middle| H \middle| \phi_\nu \right\rangle, \\ S_{\mu\nu} &= \left\langle \phi_\mu \middle| \phi_\nu \right\rangle. \end{aligned} \tag{3.3.16}$$

and the Kohn-Sham one-electron Hamiltonian is given by eq.(3.3.11).

The interatomic forces for MD applications can be derived from a calculation of the gradients of the total energy at the atom sites, divided into an electronic and a nuclear part

$$\mathbf{F}_J = -\nabla E = \mathbf{F}_{J,e} - \sum_{J\neq I} \nabla \frac{Z_I Z_J}{\left|\mathbf{R}_I - \mathbf{R}_J\right|} \tag{3.3.17}$$

where the last term is the repulsion due to the other nuclei. $\mathbf{F}_{J,e}$ can be written as a sum of orbital contributions

$$\mathbf{F}_{J,e} = \sum_i n_i \sum_\mu \sum_\nu c_{\mu i} c_{\nu i}\left(-\nabla h_{\mu\nu} + \varepsilon_i \nabla S_{\mu\nu} + \nabla V_{\mu\nu}\right) \tag{3.3.18}$$

where n_i is the occupation number of the *i*th orbital, ε_i are the orbital energy eigenvalues and $V_{\mu\nu}$ are the matrix elements of the last two terms in eq.(3.3.11).

It has been shown [74, 75] that the nuclear repulsion force from eq.(3.3.17) and the contribution arising from the last term in eq.(3.3.18) largely compensate each other and these two terms can be described together as an empirical repulsive force parameterised by a potential of the form

$$U(R)=\begin{cases} a(R-R_0)^2, & R\leq R_0 \\ 0, & R>R_0 \end{cases} \qquad (3.3.19)$$

where the parameter a is the vibrational frequency and R_0 the equilibrium bond distance for the C_2 molecule.

This model is thus a hybrid of the DF methodology together with empirical potentials. The Newton equations of motion are solved simultaneously with the approximate Kohn-Sham equations that are solved self-consistently.

3.3 Non-adiabatic MD Simulations

The above MD treatments can only describe systems in the electronic ground state. This is not such a problem for low energy collisions where the excitation is predominantly via vibrational excitation but it is a problem for high energy collisions and electron impact collisions (as well as laser excitation) where one can expect a substantial amount of electronic excitation. Time-dependent density functional methods have been developed in recent years that can describe photoabsorption spectra and excitation energies of atoms and small molecules [76], however, it is still computationally very challenging to apply such methods to collisions involving molecules as large as fullerenes. Kunert and Schmidt have applied a model to ion fullerene collisions where the electronic and vibrational degrees of freedom are treated simultaneously and self-consistently by combining time-dependent density functional theory with classical MD [54, 77]. They studied the energy loss for various ion - C_{60} collisions (H^+, C^+, Ar^+) as a function of impact velocity in the range 0.01-0.5 a.u. The relative contributions (electronic and vibrational) to the total excitation energy were shown to depend strongly on ion mass and impact velocity up to a value of 0.25 a.u. At higher velocities the excitation energies were predicted to be velocity-independent, in contradiction with the "stopping power" predictions discussed in section 3.1.1. Results of Ar^+-C_{60} collisions are shown in Fig. 3.12 for two different impact velocities (0.02 and 0.45 a.u.). At the lower velocity, the excitation is predominantly vibrational and leads to a very rapid (< 100 fs) break-up of the carbon cage. At the higher velocity, no momentum is transferred to the cage on impact but the fullerene becomes electronically excited (with approximately the same total excitation energy as for the low velocity impact). The cage starts to visibly vibrate after about 45 fs but fragments on a much longer time scale than in the first example. This type of non-adiabatic MD simulation is potentially a very powerful tool to predict and demonstrate the dynamics of collisions involving large molecules and clusters and it is to be hoped and expected that further development and application of such models will occur in the near future.

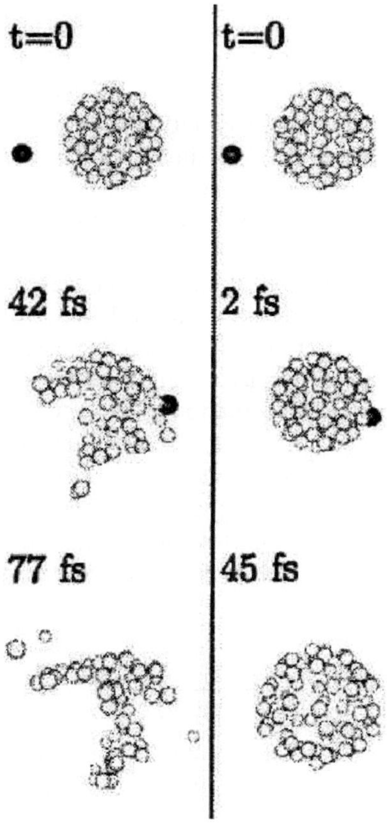

Figure 3.12: Snapshots of the reaction dynamics as a function of time for Ar^+-C_{60} collisions calculated using a combined time-dependent DFT, classical MD approach. (a) b = 2 a.u., v = 0.02 a.u. Excitation is predominantly vibrational (b) b = 2 a.u., v = 0.45 a.u. Excitation is predominantly electronic. Although the total excitation energy is approximately the same, the nature of the excitation strongly influences the time scales and mechanisms of fragmentation. Adapted from [54].

4. CHARGE TRANSFER

Many experimental studies have been made of charge transfer collisions involving fullerenes (see Chaps. 4,6 and 9). In order to interpret the results of experiments, relatively simple classical or semi-classical models are invoked that were originally developed to describe atom-ion collisions.

4.1 Classical "Over-the-Barrier" Model

The most extensive charge transfer studies in the literature involve slow collisions between highly charged ions and neutral fullerene targets (Chap. 6). In this case the classical "over-the-barrier" model is often used as a basis for models used to describe and interpret the experimental results [12, 78]. This approach considers the classical situation of a slow ion with charge q (treated as a point charge) at a distance R outside an infinitely conducting sphere with radius r_S, Fig. 3.13. The potential outside the sphere consists of two parts: one Coulombic term from the ion itself and one part from the image charge induced by the ion. If an electron tries to leave the sphere it will feel forces from the ion and its induced dipole, from the dipole that it induces itself and from the electrically charged sphere. The sphere surface must remain an equipotential surface for all distances R between the ion and the centre of the sphere. This criterion gives a unique solution for the magnitude $q' = -qr_S / R$ and position $R' = r_S^2 / R$ of the image charge. In order to conserve the total charge, an equally large positive charge $-q' = +qr_S / R$ must be placed at the centre of the sphere. The energy required to ionise the sphere in the presence of the projectile ion is shifted due to the shift in the electric potential, ϕ. For the point on the surface of the sphere closest to the projectile ion this is given as

$$\begin{aligned}\phi &= \frac{q}{R-r_S} + \frac{q'}{r_S - R'} - \frac{q'}{r_S} \\ &= \frac{q}{R-r_S} - \frac{q}{R-r_S} + \frac{qr_S}{r_S R} = \frac{q}{R}\end{aligned} \qquad (3.4.1)$$

so that the required ionisation energy, IP* = IP+q/R.

The critical distance between target and projectile at which the first electron may leave the sphere is given by the classical over-the-barrier criterion $V_e^1(\max) = IP_1^*$ where $V_e^1(\max)$ is the maximum of the potential seen by the electron when it moves towards the projectile ion. It can be written in terms of the electron's interactions with the projectile charge, the

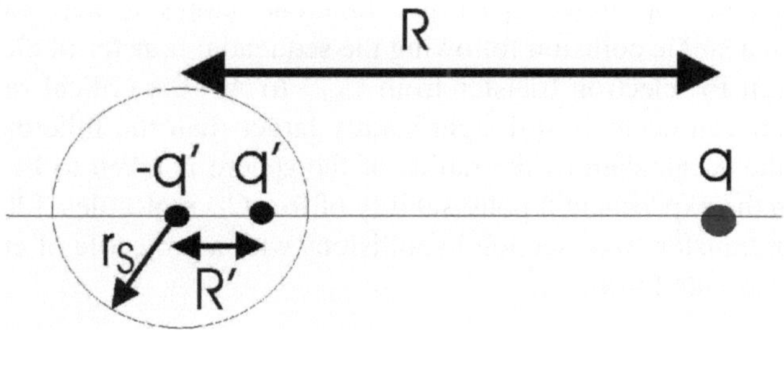

Figure 3.13. An ion of charge q at a distance R outside an infinitely conducting neutral sphere of radius r_S. The position of the image charge $q' = -r_S q/R$ is given by $R'=r_S^2/R$. An equally large positive charge $-q'$ is positioned at the centre of the sphere. Adapted from [12]

induced charges and the charge left behind on the sphere (represented by a +1 charge at the sphere centre)

$$V_e^1 = -\frac{q}{R-x} - \frac{q'}{x-R'} + \frac{q'-1}{x} + \frac{1}{2}\left(\frac{r_S}{x^2} - \frac{r_S}{x^2 - r_S^2}\right)$$
$$= -\frac{q}{R-x} + \frac{r_S q}{Rx - r_S^2} - \frac{r_S q + R}{Rx} + \frac{1}{2}\left(\frac{r_S}{x^2} - \frac{r_S}{x^2 - r_S^2}\right). \quad (3.4.2)$$

The last two terms in eq. (3.4.2) are due to the electron's interaction with its self-induced image charges. One can easily extend this to the sequential transfer of electrons. In this case the potential energy for transfer of the ith electron from an *(i-1)+* charged target to a *(q-i+1)+* charged projectile is

$$V_e^i = -\frac{(q-i+1)}{R-x} + \frac{r_S(q-i+1)}{Rx - r_S^2} - \frac{r_S(q-i+1)+iR}{Rx} + \frac{1}{2}\left(\frac{r_S}{x^2} - \frac{r_S}{x^2 - r_S^2}\right). \quad (3.4.3)$$

For electron transfer to occur this should be equal to the shifted ionisation energy for the removal of an electron from the *(i-1)+* target ion, $IP_i^* = IP_i + (q-i+1)/R$.

The potentials that fulfill the over-the-barrier criterion for transfer of electrons between a C_{60} and an Ar^{8+} projectile ion are illustrated in Fig. 3.14. The full line illustrates the situation for transfer of one electron from the neutral fullerene to the projectile ion. The critical distance for this is approximately 26 a_0. The dashed and dotted lines show the situation for transfer of one electron from the singly charged target to the 7+ projectile and from the 7+ charged target to the 1+ charged projectile ion, respectively. As the projectile ion moves past the fullerene target it may be fully neutralised in a single collision following the sequential transfer of electrons. Note that even for electron transfer from $C_{60}{}^{7+}$ to Ar^+ the critical radius at which transfer can occur is still significantly larger than the fullerene cage radius. For these calculations the radius of the sphere is taken to be 8.2 a_0, derived from the experimental polarisability of free C_{60} molecules [20]. The total electron transfer cross section in collisions with a projectile of charge q can then be estimated from

$$\sigma_{tot}^{q} = \pi R_1(q)^2 \, . \tag{3.4.4}$$

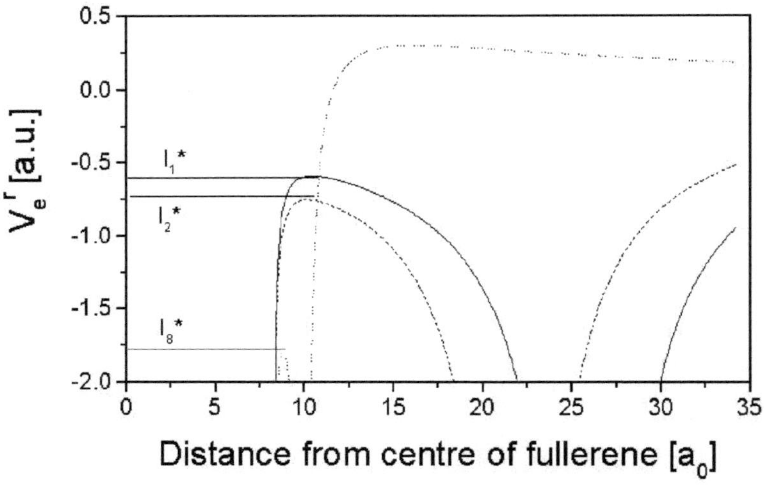

Figure 3.14. Potential energy seen by the active electron in $Ar^{8+} + C_{60}$ collisions at the critical distances for charge transfer where the height of the potential barrier matches the shifted ionisation potential I^*. Full line: transfer of the first electron from the neutral sphere to Ar^{8+}. Dashed line: transfer of the second electron from $C_{60}{}^+$ to Ar^{7+}. Dotted line: transfer of the eighth electron from $C_{60}{}^{7+}$ to Ar^+.

The above version of the over-the-barrier model assumes that the fullerene behaves like an infinitely conducting sphere. In other words, the charge mobility is assumed to be so high that all effects of localisation of individual charge carriers are averaged out. Another approach has also been applied to fullerene collisions [79-81]. In this second approach the positive holes left behind on the fullerene cage are assumed to be movable on the molecular surface in a localised fashion. During the time it takes for the electron to move from the sphere to the ion, the hole is assumed to be localised closest to the projectile. When the electron has reached the ion, the hole moves towards the position that minimises its potential energy. In this case the energy needed to ionise a neutral sphere is $IP_1^* = IP_1 + q/(R-a)$. It is more complicated to calculate the electronic potential with this model, especially if more than 2 electrons are involved. The results are not too different from the infinitely conducting sphere model. The discrepancies are larger for the transfer of a small number of electrons but as the number of holes on the sphere surface increases, the electric field outside the sphere approaches that of the infinitely conducting sphere model. For typical projectile velocities, the times between the successive critical distances are in the range 0.1-1 fs which implies that the hole re-arrangement times must be on the order of 0.1 fs or faster.

4.2 Rapp and Francis Model

The Rapp and Francis model [82] provides a semi-quantitative calculation of the collision energy dependence of charge transfer in atom-ion collisions

$$A^+ + B \rightarrow A + B^+ + \Delta E \qquad (3.4.5)$$

where ΔE is the difference in ionisation potentials of A and B. This has been applied to charge transfer collisions between neutral fullerenes and fullerene ions, discussed in Chap. 9. The method is applicable in the collision velocity regime where the semi-classical impact parameter method is applicable (roughly $(10^3/\mu^{1/2})\text{ms}^{-1} < \text{v} < 10^6\ \text{ms}^{-1}$, where μ is the reduced mass in atomic units). If one first considers symmetric charge transfer, $A^+ + A \rightarrow A + A^+$, i.e. $\Delta E = 0$, the probability of charge transfer in a collision with velocity v and impact parameter b is

$$P(b,\text{v}) = \sin^2\left[\int_{-\infty}^{\infty} \frac{(E_a - E_s)}{2\hbar \text{v}} dx\right] \qquad (3.4.6)$$

This expression is obtained from considering the A_2^+ collision complex formed during the collision as a one-electron problem. The nonstationary state (representing the collision) can be expressed in terms of symmetric and antisymmetric stationary states with energies E_a and E_s, respectively. The cross section for charge transfer is then calculated from

$$\sigma(\mathrm{v})=2\pi\int_0^\infty P(b,\mathrm{v})bdb \tag{3.4.7}$$

For ion-atom collisions this can be calculated by using the wavefunctions of the molecular ion. For more complex reactants, including fullerenes, this is not feasible and Rapp and Francis adopted an approach using semi-empirical orbitals. For R/a_0 >1, with R the internuclear separation, this becomes, in units of eV [83]

$$(E_a - E_s)=4(IP)(R/a_0)\exp\left(-\sqrt{(IP)/13.6}\,(R/a_0)\right) \tag{3.4.8}$$

where IP is the ionisation potential of atom A. Combining this with eq. (3.4.6) one can obtain an approximation to the charge transfer probability [82]

$$P_0(b,\mathrm{v}) \quad \sin^2\left[\sqrt{\frac{2\pi}{\gamma a_0}}\frac{IP}{\hbar v}b^{3/2}\left(1+\frac{a_0}{b\gamma}\right)\exp\left(-\frac{\gamma b}{a_0}\right)\right] \tag{3.4.9}$$

with $\gamma=\sqrt{IP/13.6}$. Over a fairly large velocity range this leads to a charge transfer cross section with the simple form

$$\sigma^{1/2}=-k_1\ln\mathrm{v}+k_2 \tag{3.4.10}$$

where

$$k_1=-\sqrt{\frac{\pi}{2}}\frac{a_0}{\gamma} \tag{3.4.11}$$

and k_2 is a rather complicated expression, depending only on IP and a velocity-averaged impact parameter for which $P(b,\mathrm{v})=0.25$ [82].

For asymmetric charge transfer (eq. (3.4.5)) we consider that angular momentum and spin are conserved during the charge transfer reaction. This is justifiable since charge transfer occurs at fairly large impact parameters for collisions in the "intermediate" velocity range where the model is applicable. In this case we only need to consider charge transfer between the two electronic states of AB^+ with identical symmetry which go to (A^+ + B) and (A + B^+) as asymptotic forms. This is known as a "two-state approximation", it implies that charge transfer involving other asymptotic states with a high ΔE is disregarded. The fraction of states that allow charge transfer (i.e. have identical symmetry to the initial collision partners) is denoted as *f*, the statistical weight factor. The calculation of the charge transfer cross section is quantum mechanical. The LCAO ansatz is used to obtain the total wave function for the active electron in the combined field of A^+ and B^+.

$$\psi = c_A(t)\phi_A(r_A)\exp(-i\omega_A t) + c_B(t)\phi_B(r_B)\exp(-i\omega_B t) \tag{3.4.12}$$

where the *c*'s are time-dependent coefficients of the expansion, ϕ_A and ϕ_B the atomic orbitals for the electron on on nuclei A^+ and B^+, r_A and r_B are the distances of the electron from A^+ and B^+, $\omega_i = \varepsilon_i/\hbar$ and ε_A *and* ε_B are the ionisation potentials of A and B in states ϕ_A and ϕ_B, respectively. These states are solutions of the atomic Schrödinger equations

$$\left[\left(-\frac{\hbar^2}{2m_e}\right)\nabla^2 + V_j(r_j)\right]\phi_j = \varepsilon_j\phi_j \tag{3.4.13}$$

in which *j* can be either A or B and V_j are the effective potentials binding the electron to the nuclei. The time-dependent coefficients c_j are then obtained by solving the time-dependent Schrödinger equation

$$\left[\left(-\frac{\hbar^2}{2m_e}\right)\nabla^2 + V_A(r_A) + V_B(r_B)\right]\psi = \hbar i\frac{\partial\psi}{\partial t} \tag{3.4.14}$$

Eqs. 3.4.13 and 3.4.14 thus provide a system of equations for the coefficients c_j and their time derivatives. The reaction probability, $P(b,\mathrm{v})$ is then given by

$$P(b,\mathrm{v}) = \left|c_A(t\to\infty)\right| f\ . \tag{3.4.15}$$

After a number of rearrangements and approximations that will not be detailed here [82], the charge transfer cross section can be obtained from

$$\sigma(\mathrm{v}) = \frac{1}{2}f\int_0^{b_1}\mathrm{sech}^2\left[\frac{\omega}{\mathrm{v}}\sqrt{\frac{a_0\pi b}{2\gamma}}\right]2\pi b\,db \tag{3.4.16}$$

The impact parameter b_1 is obtained from the relationship

$$\mathrm{sech}^2\left[\frac{\omega}{\mathrm{v}}\sqrt{\frac{a_0\pi b}{2\gamma}}\right] = 4P_0(b_1,\mathrm{v}) \tag{3.4.17}$$

where P_0 is obtained from eq. 3.4.9 and $\omega = \Delta E / \hbar$.

Examples of the application of this model to charge transfer collisions involving fullerenes are discussed in Chap. 9.

REFERENCES

1 M. A. D. Fluendy and K. P. Lawley, *Chemical Applications of Molecular Beam Scattering* (Chapman and Hall Ltd., New York, 1973).

2 R. D. Levine and R. B. Bernstein, *Molecular Reaction Dynamics and Chemical Reactivity* (Oxford University Press, New York, Oxford, 1987).

3 E. W. McDaniel, J. B. A. Mitchell and M. E. Rudd, *Atomic Collisions* (John Wiley and Sons, Inc., 1993).

4 B. H. Bransden and M. R. C. McDowell, *The theory of ion-atomic collisions* (Oxford Science Publ. Clarendon Press, Oxford, 1992).

5 E. E. B. Campbell, R. Ehlich, G. Heusler, O. Knospe and H. Sprang, Chem. Phys. **239**, 299 (1998).

6 J. Lindhard, V. Nielsen and M. Scharff, Mat. Fys. Medd. Dan. Vid. Selsk. **36**, 1 (1968).

7 J. Lindhard, V. Nielsen, M. Scharff and P. V. Thomsen, Mat. Fys. Medd. Dan. Vid. Selsk. **33** (1963).

8 J. F. Ziegler, J. P. Biersack and U. Littmark, *The Stopping and Ranges of Ions in Solids* (Pergamon Press, New York, 1985).

9 J. F. Ziegler and J. P. Biersack, TRIM-code (1991).

10 A. Reinköster, U. Werner and H. O. Lutz, Europhys. Lett. **43**, 653 (1998).

11 M. C. Larsen, P. Hvelplund, M. O. Larsson and H. Shen, Eur. Phys. J. D **5**, 283 (1999).

12 H. Cederquist, A. Fardi, K. Haghighat, A. Langereis, H. T. Schmidt, S. H. Schwartz, J. C. Levin, *et al.*, Phys. Rev. A **61**, 022712 (2000).

13 J. Opitz, H. Lebius, S. Tomita, B. A. Huber, P. Moretto-Capelle, D. Bordenave-Montesquieu, A. Bordenave-Montesquieu, *et al.*, Phys. Rev. A **62**, 022705 (2000).

14 A. V. Glotov and E. E. B. Campbell, Chem. Phys. Lett. **327**, 61 (2000).

15 B. Brunetti, P. Candori, S. Falcinelli, F. Vecchiocattivi, A. Sassara and M. Chergui, J. Phys. Chem. A **104**, 5942 (2000).

16 L. A. Girifalco, J. Chem. Phys. **95**, 5370 (1991).

17 L. A. Girifalco, J. Phys. Chem. **96**, 858 (1992).

18 J. P. K. Doye and D. J. Wales, Chem. Phys. Lett. **262**, 167 (1996).

19 J.P.K.Doye, A. Dullweber and D. J. Wales, Chem. Phys. Lett. **269**, 408 (1997).

20 A. A. Scheidemann, V. V. Kresin and W. D. Knight, Phys. Rev. A **49**, R4293 (1994).

21 R. Antoine, P. Dugourd, D. Rayane, E. Benichou, M. Broyer, F. Chandezon and C. Guet, J. Chem. Phys. **110**, 9771 (1999).

22 N. Andersen, J. Broad, E. Campbell, J. Gallagher and I. V. Hertel, Phys. Rep. **278**, 107 (1997).

23 E. E. B. Campbell and R. D. Levine, Annu. Rev. Phys. Chem. **51**, 65 (2000).

24 S. Matt, O. Echt, M. Sonderegger, R. David, P. Scheier, J. Laskin, C. Lifshitz, *et al.*, Chem. Phys. Lett. **303**, 379 (1999).

25 C. Lifshitz, Int. J. Mass Spec. **198**, 1 (2000).

26 C. Lifshitz, Chem. Soc. Rev. **30**, 186 (2001).

27 A. D. Boese and G. E. Scuseria, Chem. Phys. Lett. **294**, 233 (1998).

28 K. Hansen and O. Echt, Phys. Rev. Lett. **78**, 2337 (1997).

29 J. Laskin, B. Hadas, T. D. Märk and C. Lifshitz, Int. J. Mass Spectrom. Ion Proc. **177**, L9 (1998).
30 S. Tomita, J. U. Andersen, C. Gottrup, P. Hvelplund and U. V. Pedersen, Phys. Rev. Lett. **87**, 073401 (2001).
31 S. Diaz-Tendero, M. Alcami and F. Martin, J. Chem. Phys. **119**, 5545 (2003).
32 S. Tomita, J. U. Andersen, K. Hansen and P. Hvelplund, , submitted (2003).
33 P. E. Barran., S. Firth, A. J. Stace, H. W. Kroto, K. Hansen and E. E. B. Campbell, Int. J. Mass Spectrom. Ion Proc. **167/168**, 127 (1997).
34 J. A. Zimmerman, J. R. Eyler, S. B. H. Bach and S. W. McElvany, J. Chem. Phys, **94**, 3556 (1991).
35 H. Steger, J. de Vries, W. Kamke, I. V. Hertel and T. Drewello, Chem. Phys. Lett. **194**, 452 (1992).
36 P. Scheier, B. Dünser, R. Wörgötter, M. Lezius, R. Robl and T. D. Märk, Int. J. Mass Spec. Ion Proc. **138**, 77 (1994).
37 G. Javahery, H. Wincel, S. Petrie and K. D. Bohme, Chem. Phys. Lett. **204**, 467 (1993).
38 C. Lifshitz, M. Iraqui, T. Peres and J. E. Fischer, Rapid Commun. Mass Spectrom. **5**, 238 (1991).
39 H. Steger, J. Holzapfel, A. Hielscher, W. Kamke and I. V. Hertel, Chem. Phys. Lett. **234**, 455 (1995).
40 R. Völpel, G. Hofmann, M. Steidl, M. Stenke, M. Schlapp, R. Trassl and E. Salzborn, Phys. Rev. Lett. **71**, 3439 (1993).
41 G. Seifert, R. Gutierrez and R. Schmidt, Phys. Lett. A **211**, 357 (1996).
42 S. Matt, O. Echt, R. Wörgötter, V. Grill, P. Scheier, C. Lifshitz and T. D. Märk, Chem. Phys. Lett. **264**, 149 (1997).
43 J. U. Andersen, E. C. Bonderup and K. Hansen, J. Phys. B **35**, R1 (2001).
44 K. Hansen, Philos Mag. B **79**, 1413 (1999).
45 C. E. Klots, Chem. Phys. Lett. **186**, 73 (1991).
46 S. G. Kim and D. Tománek, Phys. Rev. Lett. **72**, 2418 (1994).
47 H. Hohmann, R. Ehlich, S. Furrer, O. Kittelmann, J. Ringling and E. E. B. Campbell, Z. Phys. D **33**, 143 (1995).
48 J. Silberstein and R. D. Levine, J. Chem. Phys. **75**, 5735 (1981).
49 T. Raz, U. Even and R. D. Levine, J. Chem. Phys. **103**, 5394 (1995).
50 E. E. B. Campbell, T. Raz and R. D. Levine, Chem. Phys. Lett. **253**, 261 (1996).
51 A. Lassesson, A. Gromov, K. Mehlig, A. Taninaka, H. Shinohara and E. E. B. Campbell, J. Chem. Phys. **119**, 5591 (2003).
52 E. E. B. Campbell, A. Glotov, A. Lassesson and R. D. Levine, C.R. Physique **3**, 341 (2002).
53 D. R. Stull and H. Prophet, (Natl. Bur. Stand., Washington DC, 1970).
54 T. Kunert and R. Schmidt, Phys. Rev. Lett. **86**, 5258 (2001).
55 B. Torralva, T. A. Niehaus, M. Elstner, S. Suhai, T. Frauenheim and R. E. Allen, Phys. Rev. B **64**, 153105 (2001).
56 D. W. Heermann, *Computational Simulation Methods in Theoretical Physics* (Springer, Berlin, 1990).
57 L. Verlet, Phys. Rev. **159**, 98 (1967).
58 T. Frauenheim, D. Porezag, T. Köhler and F. Weich, in *Computational Physics*, edited by K. H. Hoffmann and M. Schreiber (Springer, Berlin, 1996), p. 294.
59 J. Tersoff, Phys. Rev. Lett. **61**, 2879 (1988).
60 D. W. Brenner, Phys. Rev. B **42**, 9458 (1990).
61 D. W. Brenner, Phys. Rev. B **46**, 1948 (1990).
62 R. Ehlich, E. E. B. Campbell, O. Knospe and R. Schmidt, Z. Phys. D **28**, 153 (1993).

63 D. H. Robertson, D. W. Brenner and C. T. White, J. Phys. Chem. **99**, 15721 (1995).
64 R. C. Mowrey, M. M. Ross and J. H. Callahan, J. Phys. Chem. **96**, 4755 (1992).
65 G. Moliere, Z. Naturf. **2A**, 133 (1947).
66 A. V. Glotov, O. Knospe, R. Schmidt and E. E. B. Campbell, Eur. J. Phys. D **16**, 333 (2001).
67 R. Ehlich, M. Westerburg and E. E. B. Campbell, J. Chem. Phys. **104**, 1900 (1996).
68 P. Hohenberg and W. Kohn, Phys. Rev. **136**, B864 (1964).
69 W. Kohn and L. J. Sham, Phys. Rev. **140**, A1133 (1965).
70 R. Ehlich, PhD Thesis (Universität Freiburg) (1994).
71 K. Ohno, Y. Maruyama, K. Esfarjani, Y. Kawazoe, N. Sato, R. Hatakeyama, T. Hirata, *et al.*, Phys. Rev. Lett. **76**, 3590 (1996).
72 B. L. Zhang, C. Z. Wang, C. T. Chan and K. M. Ho, J. Phys. Chem. **97**, 3134 (1993).
73 O. Knospe and R. Schmidt, in *Theory of Atomic and Molecular Clusters with a Glimpse at Experiments*, edited by J. Jellinek (Springer Verlag, 1999), p. 111.
74 G. Seifert, H. Eschrig and W. Bieger, Z. Phys. Chem. (Leipzig) **267**, 529 (1986).
75 G. Seifert and H. Eschrig, Phys. Stat. Sol. (B) **127**, 573 (1985).
76 E. Runge and E. K. U. Gross, Phys. Rev. Lett. **52**, 997 (1984).
77 U. Saalmann and R. Schmidt, Phys. Rev. Lett. **80**, 3213 (1998).
78 A. Bárány and C. J. Setterlind, Nucl. Instrum. Methods Phys. Res. B **98**, 184 (1995).
79 S. Petrie, J. Wang and D. K. Bohme, Chem. Phys. Lett. **204**, 473 (1993).
80 D. B. Cameron and J. H. Parks, Chem. Phys. Lett. **272**, 18 (1997).
81 H. Shen, P. Hvelplund, D. Mathur, A. Bárány, H. Cederquist and N. Selberg, Phys. Rev. A **52**, 3847 (1995).
82 D. Rapp and W. E. Francis, J. Chem. Phys. **37**, 2631 (1962).
83 D. P. Dewangan, J. Phys. B: Atom. Molec. Phys. **6**, L20 (1973).

Chapter 4

CHEMICALLY REACTIVE AND THERMAL ENERGY COLLISIONS

In this chapter we will review collision experiments at thermal or quasi-thermal energies that have been used to study mainly charge transfer and collision reactions. The first part of the chapter will consider the wide range of reactions that have been studied using the SIFT method (selected ion flow tube). We will also touch on some complementary techniques applied to chemical reaction studies involving fullerenes, also at higher collision energies. In the second part of the chapter we will discuss Penning ionisation studies of collisions between neutral fullerenes and electronically excited rare gas atoms.

1. LOW ENERGY ADDITION AND CHARGE TRANSFER REACTIONS WITH FULLERENES

1.1 Addition Reactions

Low energy addition reactions and charge transfer involving fullerene ions have been extensively studied by Bohme and co-workers [1-4] using the SIFT technique (see Fig. 2.8). The reactions of singly, doubly and triply charged C_{60} with a wide range of collision partners were studied at room temperature (294 ± 2 K) and at a helium bath pressure of 0.35 ± 0.01 Torr [2].

The positive charge on the fullerene can be regarded as being delocalised over the carbon surface. The interaction with an incoming molecule will then localise the charge as a consequence of the electrostatic interaction between the charge and the induced or permanent dipole of the molecule. Chemical addition can then occur at the site of the charge. However, the carbon atom on the cage that undergoes the chemical bond has to experience a change in hybridisation from sp^2 to sp^3 that implies that exothermic covalent bond formation may involve an activation barrier. The singly-charged fullerene cation is in fact rather unreactive towards many molecules [2]. The reactions that have been observed in SIFT experiments are summarised in Fig. 4.1. As the charge on the cation increases, it becomes easier to derivatise the cage and considerably more reaction products could be observed in collisions with C_{60}^{2+} than with C_{60}^{+} [2]. This is presumably because the stronger electrostatic interaction between C_{60}^{2+} and an incoming molecule can help to overcome the activation barrier associated with the hybridisation of the bonding site. As the positive charge on the fullerene increases further, the ionisation energy of the fullerene increases and electron transfer becomes more exothermic and kinetically more favourable, in spite of the activation barrier that arises from the production of two charged products. Electron transfer therefore becomes more likely as a competing or dominant reaction channel. For this reason the number of observed addition reactions with a triply charged fullerene cation is lower than that for a doubly charged cation [2].

Interesting trends are seen in the reactions of fullerene cations with ammonia and amines [5]. Ammonia adds only very slowly to the singly charged cation but the rate of addition approaches the collision rate with increasing alkyl substitution, Table 4.1. The reactions occur exclusively by adduct formation

$$C_{60}^{\dot{+}} + M \rightarrow C_{60}M^{\,+} \tag{4.1.1}$$

where the dot indicates a free radical species. The adduct formation is attributed to the formation of a covalent C-N bond with the fullerene cation through donation of an electron pair from the nitrogen atom to the site of charge concentration on the surface of the fullerene, as illustrated in Fig. 4.2. This requires a transition from a reactant ion with extreme charge delocalisation to a product ion where the charge is extremely localised on the nitrogen atom. This formation mechanism is quite different to the proposed mechanism for amine addition to neutral C_{60} in solution which involves two steps: single electron transfer to a π bond followed by covalent bond formation and proton transfer [6].

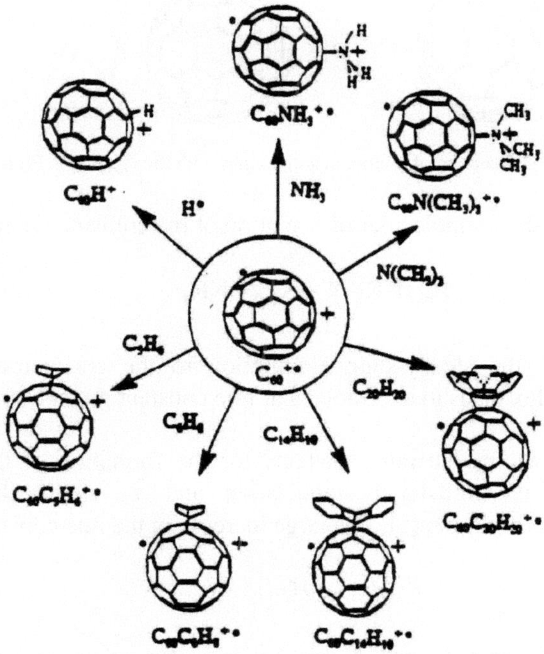

Figure 4.1. Overview of addition reactions of C_{60}^+ observed with a SIFT apparatus at room temperature and at a helium bath gas pressure of 0.35 Torr. From [2].

Table 4-1. Products and rate coefficients for reactions of C_{60}^+ with ammonia and amines at 294 K in helium, determined using the SIFT technique. From [5]. The rate coefficients are given in units of $10^{-9}cm^3molecule^{-1}s^{-1}$. k_C is the collision rate coefficient.

neutral reactant	product	k_{obs}	k_c
NH_3	$C_{60}NH_3^+$	< 0.001	1.69
CH_3NH_2	$C_{60}CH_3NH_2^+$	0.015	1.44
$CH_3CH_2NH_2$	$C_{60}CH_3CH_2NH_2^+$	0.050	1.34
$(CH_3)_2NH$	$C_{60}(CH_3)_2NH^+$	0.85	1.22
$(CH_3)_3N$	$C_{60}(CH_3)_3N^+$	1.0	1.05

The observed increasing reaction rate with increasing degree of aliphatic substitution of the amine can be explained by considering the association to occur in two steps. Firstly, a collision complex is formed which is stabilised by colllision with the He atoms of the buffer gas. The collisional stabilisation

$$\left(C_{60}NR_3^+\right)^* + He \rightarrow C_{60}NR_3^+ + He^* \qquad (4.1.2)$$

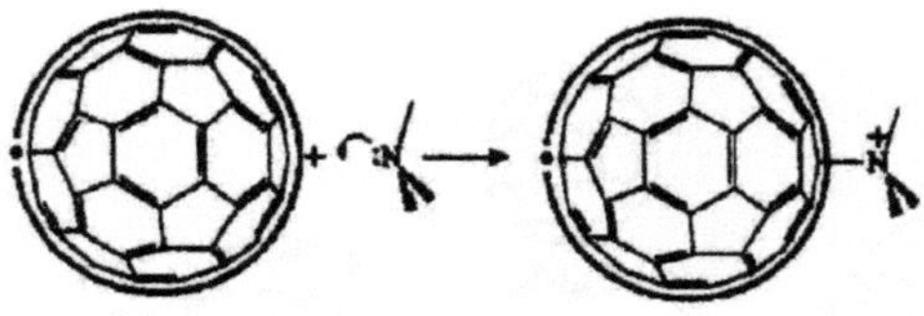

(2)

Figure 4.2. Reaction of ammonia and amines with the C_{60}^+ cage. From [5].

competes with the unimolecular dissociation of the collision complex

$$(C_{60}NR_3^+)^* \rightarrow C_{60}^+ + NR_3 \qquad (4.1.3)$$

Application of the steady-state assumption to the concentration of the collision complex leads to a termolecular rate constant given by

where k_C and k_S are the rate constants for the formation of the collision complex and its collisional stabilisation and τ_D is the lifetime for unimolecular dissociation [5]. The large increase in the rate constant is due

$$k(3) = k_C k_S \tau_D \qquad (4.1.4)$$

primarily to τ_D since k_C and k_S are not expected to be very sensitive to aliphatic substitution. Various statistical theories can be used to provide an estimate of τ_D of which the simplest is [7]

$$\tau_D = \tau_0 \left(\frac{(D + 3RT)}{3RT} \right)^{S-1} \qquad (4.1.5)$$

where τ_0 is a collision lifetime corresponding to the frequency of vibration along the reaction coordinate, D is the dissociation energy of the collision complex, S is the number of effective degrees of freedom available in the complex for energy transfer and R is the gas constant. Although C_{60} has 60 atoms, it has no internal modes of rotation and, given its highly rigid structure, the vibrational modes are widely space and will not allow for a great deal of energy dispersion within the collision complex under the experimental conditions. Trimethylamine contains much fewer degrees of freedom than C_{60} however, the methyl groups have rotational modes that have a much higher energy density than the vibrational modes of C_{60}. The bending vibrational modes of trimethylamine will also have a relatively high energy density, suggesting that they may also aid in energy dispersal within the collision complex and thus prolongation of the complex lifetime.

The doubly charged fullerene cations show quite different behaviour, Table 4.2. In this case adduct formation competes with charge transfer. Charge transfer is endothermic for the reactions with the singly charged cation but exothermic for reactions with the doubly charged cation. However, Coulomb repulsion between the singly charged cations produced in a charge transfer reaction introduces a barrier to charge transfer. Therefore charge transfer should only be observable at room temperature for reactions that are sufficiently exothermic to overcome this barrier [8]. The ionisation potential of C_{60}^+ is 11.4 eV which implies that charge transfer with all of the amines in Table 4.2 is exothermic by more than 2 eV. Charge transfer is

Table 4-2. Products and rate coefficients for reactions of C_{60}^{2+} with ammonia and amines at 294 K, determined using the SIFT technique. Rate coefficients in units of 10^{-9} $cm^3 molecule^{-1} s^{-1}$. From [5].

neutral reactant	product	branching ratio	k_{obs}	k_c
NH_3	$C_{60}NH_3^{2+}$	1.00	1.2	3.37
CH_3NH_2	$C_{60}CH_3NH_2^{2+}$	0.90	2.6	2.88
	$C_{60}^+ + CH_3NH_2^+$	0.10		
$CH_3CH_2NH_2$	$C_{60}CH_3CH_2NH_2^{2+}$	0.70	2.6	2.68
	$C_{60}^+ + CH_3CH_2NH_2^+$	0.30		
$(CH_3)_2NH$	$C_{60}(CH_3)_2NH^{2+}$	0.10	2.9	2.44
	$C_{60}^+ + (CH_3)_2NH^+$	0.90		
$(CH_3)_3N$	$C_{60}^+ + (CH_3)_3N^+$	1.00	2.2	2.10

indeed observed to compete with association for all the amines and, according to the measured branching ratios, does so increasingly as the charge transfer reaction becomes more endothermic until it is the only observed channel in the reaction with trimethylamine which has the lowest ionisation potential (7.82 eV) of all the amines in the study. No charge transfer is observed with ammonia (IP = 10.16 eV) even although this channel is also exothermic. However, the exothermicity of the charge transfer reaction (1.23 eV) is less than the contribution to the barrier height which arises from the Coulomb repulsion [5]. The situation is illustrated schematically in Fig. 4.3.

No association reaction products were found in reactions between the amines and C_{60}^{3+}. The triply charged cation reacted exclusively by charge transfer, in accordance with the much higher ionisation energy of C_{60}^{2+}.

Triple quadrupole mass spectrometry experiments have also been applied to the study of the association reactions of fullerene ions [9]. Several factors have been shown to influence the addition reactions: the charge state and size of the fullerene ion and the size, structure and degree of unsaturation of the target.

The reactivity of fullerene anions has been studied with the flowing afterglow technique [10]. They were shown to be rather unreactive with a variety of acids but did react with NO_2.

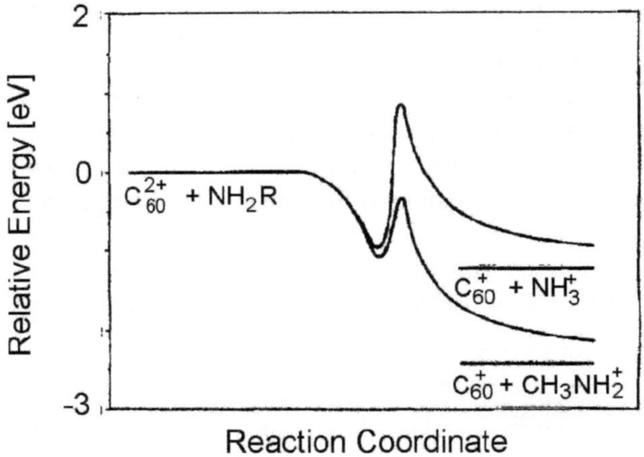

Figure 4.3. Hypothetical potential energy / reaction coordinate profile which accounts for the occurrence and nonoccurrence of charge transfer in reactions between C_{60}^{2+} and amines. The interaction between the product ions is dominated by Coulombic repulsion.

Time of flight mass spectrometry experiments, at much higher collision energies than discussed above, have also been applied to the study of fullerene reactions. Callahan et al. carried out detailed studies of the reaction products in collisions between fullerene ions and O_2 [11]. Their results shed some light on the photostability of fullerenes in air and indicated that the reaction with O_2 and O_3 (but not O) led to the disruption of the cage structure and fragmentation to produce CO, CO_2 and C_2O_3. The paper also discussed an empirical link between the extractability of macroscopic amounts of endohedral metallofullerenes and their oxidative stability.

Guided ion beam experiments were applied to the study of reactive collisions between neutral fullerenes and atomic cations. Some of these results involve the endohedral capture of the ion inside the carbon cage [12-15] and are discussed at length in Chapter 8. Three different types of bonding were studied in the experiments: the formation of weakly bound substituents attached to the carbon cage, the replacement of carbon atoms in the fullerene cage by heteroatoms and the above-mentioned formation of endohedral complexes. A summary of the results obtained for the different collision partners investigated is given in Table 4.3. One of the most interesting systems was $C^+ + C_{60}$ where evidence was found for both a weakly bound C attached to the cage and for a substitution reaction in hich the projectile ion replaced one of the carbon atoms [16]. The cross sections measured for the range of products is shown as a function of laboratory collision energy in Fig. 4.4. The C_{61}^+ product ion appears over a narrow energy window, peaking at an energy of approximately 4 eV. The most dominant process over the entire collision energy range investigated is charge transfer to produce C_{60}^+. The nature of the C_{61}^+ adduct was addressed

Table 4-3. Summary of results from reactive ion-fullerene collisions obtained by the Anderson group [17,18].

Projectile ion	low energy adduct formed?	Approx. disappearance energy of adduct [eV]	Substitutional compound formed?	Endohedral complex formed?	Efficient charge transfer?
Li^+, Na^+, K^+	no	-	no	yes	no
B^+	yes	10	no	no	yes
C^+	yes	15	yes	no	yes
N^+	no	-	yes	no	yes
O^+	yes	28	yes?	yes?	yes
F^+	no	-	no	no	yes
Ne^+	no	-	no	yes	yes
Mn^+	yes	10	no	yes	yes
Fe^+	yes	15, 40	no	?	yes
C_2^+	yes	36	yes?	no	yes
CO^+	no	-	no	no	yes
Mn_2^+	only MnC_{60}^+	13	no	only $Mn@C_{60}^+$	yes

by using a $^{13}C^+$ projectile ion and measuring the isotopomer distribution resulting from the process $[C_{61}^+]^* \rightarrow C_{60}^+ + C$. There are two possible limiting cases. If the extra carbon atom is only weakly bound to the cage then the dominant decay route will be the loss of the added C atom with no isotope scrambling. At the other limit, the C_{61}^+ could incorporate the extra atom, producing a distorted fullerene. In this case the ^{13}C atom is more likely to be retained in the subsequent decay process. Since the C_{61}^+ is internally hot, bond rearrangements could lead to a completely random distribution in the cage and therefore a small probability for losing the ^{13}C. Figure 4.5 shows the isotope distributions for C_{60}^+ and C_{58}^+ produced in the reactions between $^{13}C^+$ and C_{60} [16]. Fig. 4.5(a) shows the C_{60}^+ peak under conditions where the contribution to the signal contains both charge transfer products as well as the decay product from C_{61}^+. The pure charge transfer product will simply show the natural isotope distribution. The distribution in Fig 4.5(a) shows some evidence for retention of ^{13}C. Parts (c) and (d) of the figure show the expected isotope distributions assuming the natural ^{13}C abundance and also assuming that there is complete isotope scrambling. In Fig. 4.5(b) only forward scattered ions are detected and therefore the pure charge transfer signal is discriminated against. Under these conditions the isotope distribution clearly shifts towards a higher ^{13}C content indicating the presence of isotopic scrambling in the cage [16]. Tight binding MD simulations were able to provide good overall agreement with the experimental results [19]. However, their predictions of endohedral formation at high collision energies could not be confirmed by experiment. Indications for the possibility of endohedral formation in $C^+ + C_{60}$ collisions at high collision energies (velocity = 0.02 a.u.) were also found in nonadiabatic MD simulations [20] but such collision complexes would not survive intact for times long enough to be detected in the experiment.

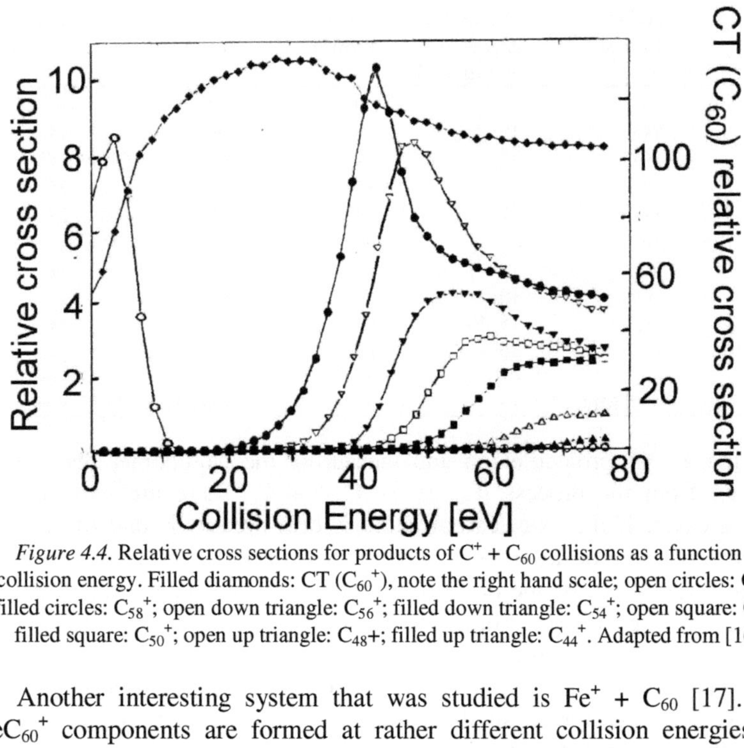

Figure 4.4. Relative cross sections for products of $C^+ + C_{60}$ collisions as a function of collision energy. Filled diamonds: CT (C_{60}^+), note the right hand scale; open circles: C_{61}^+; filled circles: C_{58}^+; open down triangle: C_{56}^+; filled down triangle: C_{54}^+; open square: C_{52}^+; filled square: C_{50}^+; open up triangle: C_{48}+; filled up triangle: C_{44}^+. Adapted from [16].

Another interesting system that was studied is $Fe^+ + C_{60}$ [17]. Two FeC_{60}^+ components are formed at rather different collision energies that appear to have different stabilities although the fragmentation patterns argue against an endohedral product. The CID dissociation of a relatively strongly exohedrally bound Fe-C_{60} complex formed by ligand exchange reactions with fullerenes had been investigated previously [21].

1.2 Charge Transfer

Low energy charge transfer reactions are closely related to the addition reactions discussed in the previous section and often provide a competing reaction channel. The most extensive low energy studies have been carried out by Bohme and collaborators [1]. Various experiments have shown that C_{60} can be chemically ionised at thermal and higher energies in single-

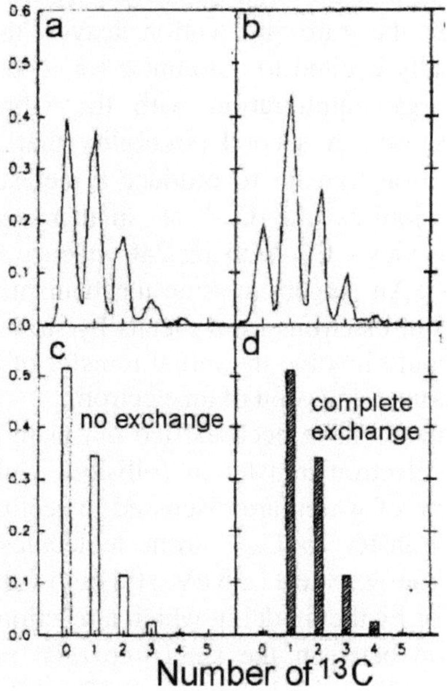

Figure 4.5. Isotopomer distributions resulting from reaction of $^{13}C^+$ with C_{60}. (a) C_{60}^+ at low collision energy and high octapole trapping voltage (b) C_{60}^+ at low collision energy with octapole set to discriminate against charge transfer products. (c) isotopomer distribution expected for no isotope scrambling (d) isotopomer distributions expected for full isotopic scrambling. Adapted from [16].

electron transfer reactions with cations that have an electron recombination energy that is greater than or equal to the ionisation potential of C_{60}. Early observations involved charge transfer between atomic metal ions and C_{60} [21,22]. Rate coefficients and branching fractions for charge transfer were not reported but Fe^+ was said to react with C_{60} predominantly via charge transfer. Reactions of C_{60} with other atomic ions having higher electron recombination energies than 7.6 eV e.g. D^+ (13.6 eV), Si^+ (8.151 eV) and the rare gas ions have been investigated in SIFT experiments at 294 K [23,24]. All of these ions were found to extract an electron from C_{60} under the conditions of the experiment.

Reactions of He^+ and Ne^+ with C_{60}, studied in SIFT experiments, produced substantial amounts of doubly charged C_{60}^{2+} in electron transfer, electron detachment reactions of the type

$$X^+ + C_{60} \rightarrow C_{60}^{2+} + X + e^- \qquad (4.1.6)$$

that competed with single electron transfer [23]. No dication formation was found for collisions with Ar^+. Several mechanisms have been proposed to describe removal of two electrons from C_{60} in its reaction with He^+ and Ne^+ [23]. One possibility is that the high exothermicity of the transfer of one electron from C_{60} to the rare gas cation leaves the fullerene cation sufficiently electronically excited to autoionise (or to undergo fast thermal ionisation before energy equilibration with the vibrational degrees of freedom, see next section). A second possibility that has been proposed involves an intial electron transfer to produce a metastable rare gas atom which then Penning ionises the C_{60}^+ at smaller separations. Another alternative mechanism views C_{60} with its 240 valence electrons to have a metalloid character [1]. An Auger emission mechanism has long been used to explain the ejection of electrons from metals by slow rare gas ions. With C_{60} this mechanism would involve the initial transfer of a deeply embedded electron followed by Auger emission of an electron.

Extensive measurements have been carried out using SIFT to determine the rate constants for electron transfer in collisions with multiply charged fullerene ions [1], some of which are discussed in section 4.1.1 above. The absence of electron transfer to C_{60}^{2+} from molecules having ionisation energies in the range from 9.58 to 11.36 eV, still below the ionisation energy of C_{60}^+, is accounted for by the model in which a reaction barrier arises from the Coulomb repulsion between the singly charged product ions of the electron transfer process, see 4.1.1 above and Fig. 4.3. This assumes that electron transfer is most likely to occur when the reactants are in close proximity and one of the charges has been localised by the electrostatic interaction between the reactants with the other charge diametrically opposed on the fullerene surface. The requirement for electron transfer from a dication to a neutral molecule, producing two monocations initially in close proximity, is that the reaction exothermicity must exceed the electrostatic potential which denotes the Coulomb repulsion between the monocations at the moment of electron transfer (see Chap. 3.4.1). The Coulomb repulsion depends on the distance between the charges. If it is assumed that the charge separation at the moment of electron transfer is equal to or greater than the diameter of the fullerene cage then the electrostatic potential is < 2.06 eV [1]. This agrees well with the experimental observations.

Single electron transfer reactions with triply charged C_{60}^{3+} have also been studied for a wide range of molecular targets [25,26]. The ionisation energy threshold for the occurrence of electron transfer to C_{60}^{3+} lies well below the thermodynamics threshold by about 4.4. eV. This is higher than the analogous differnce observed for dication reactions. Again, the explanation lies in the Coulombic repulsion since the products of single-electron transfer are both charged. In this case we have one singly and one doubly charged ion initially in close proximity to one another. The threshold for production of such an ion pair will be significantly above the threshold or production of ions at infinite separation. A barrier of 4.5 eV has been estimated from a geometrical model for the charge distribution in $C_{60}{}^{3+}$ and the effects of Coulomb repulsion between the products [27].

Double-electron transfer has been observed between $C_{60}{}^{3+}$ and polycyclic aromatic hydrocarbons (PAHs) at room temperature [28]. Both sequential and concomitant electron transfer have been proposed as classical mechanisms of double-electron transfer to $C_{60}{}^{3+}$ [28]. In the sequential mechanism, the transfer of a single electron is proposed to occur first, followed by reorganisation of the remaining two charges on C_{60} and then the transfer of the second electron (see also 3.4.1 and Fig. 3.15). This mechanism assumes that the intial three charge sites on the cage are too far apart for two electrons to be transferred at the same time to two different charge sites. The second classical mechanism proposed involves the concomitant transfer of two electrons to form an intermediate zwitterion $C_{60}{}^{2+,-}$.

2. PENNING IONISATION

Penning ionisation of C_{60}

$$RG^* + C_{60} \rightarrow RG + C_{60}^{*+} + e^- \qquad (4.2.1)$$

has been observed in flow tube studies [29] and studied in more detail by two other groups using molecular beams [30-32]. The ionisation cross section was found to rise exponentially with increasing excitation energy. The experimental results from [30,31] are shown on Fig. 4.6. The values reported by Brunetti et al. [31] were only relative but they provide a point at low excitation energy (8.32 eV) for collisions with Xe*. The two data sets agree very well where identical collision partners were used so the data of Weber et al [30] were used to normalise the data from Brunetti et al. [31]. The data has been fitted by a model in which the energy transferred in the collision was assumed to be statistically distributed among the electronic degrees of freedom [33]. The ionisation was regarded as being thermal and the total ion yield was quenched on a time scale of the order of 240 fs by coupling to the thermal sink of the nuclear degrees of freedom of the fullerene. The coupling is characterised by a time constant which, in the original study [30] was fitted to be about a picosecond but with later, improved calculations of the electronic density of states this value was

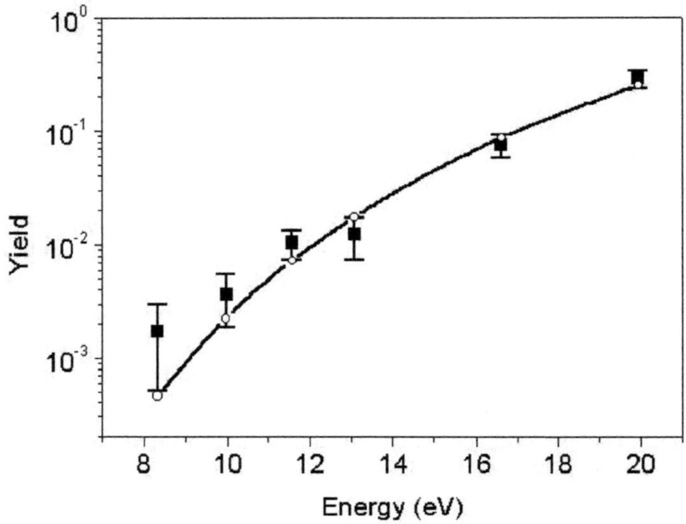

Figure 4.6. Yield of ions in Penning ionisation experiments on C_{60}. The x-axis gives the energy transferred from each excited rare gas. Data from [30,31]. The full line is a fit to the data assuming that electrons are emitted thermally before the energy can be equilibrated with the degrees of freedom [33].

revised downwards [33]. The fit of the time constant requires an expression for the electron emission rate and, since the model is statistical, also an expression for the electronic level density of C_{60} and C_{60}^+. The rate constant is given as [34,35]

$$k(E,\varepsilon)d\varepsilon = \frac{2m_e\sigma(\varepsilon)}{\pi^2\hbar^3}\varepsilon\frac{\rho_d(E-IP-\varepsilon)}{\rho_p(E)} \qquad (4.2.2)$$

where ε is the kinetic energy of the electron, m_e the mass of the electron, ρ the level density of the daughter ion and parent molecule, E the internal energy of the molecule and $\sigma(\varepsilon)$ the capture cross section for an electron in the Coulomb potential of the daughter ion. The total rate is then

$$k(E) = \int_0^E k(E,\varepsilon)d\varepsilon. \qquad (4.2.3)$$

The rate constant of eq. (1) is shown in Fig. 4.7 together with the rate constant expected when the same excitation energy is distributed over the vibrational degrees of freedom (see 3.2.1).

The model was later applied to understand the photoelectron spectra produced from the interaction of C_{60} with ultrashort laser pulses [33,36].

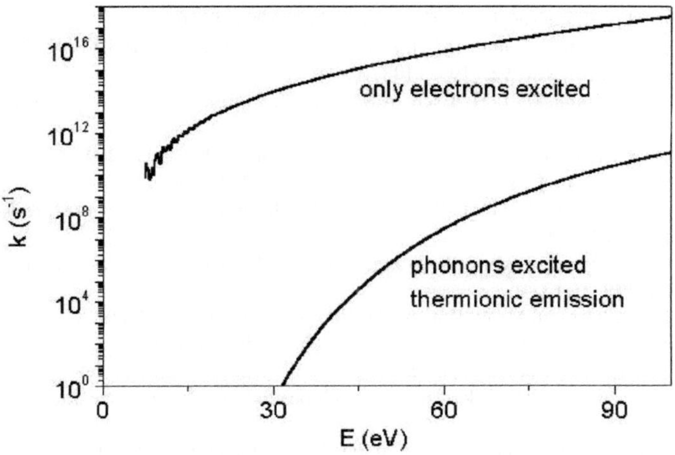

Figure 4.7. Rates for thermal electron emission from C_{60} as a function of internal energy.

REFERENCES

1 D. K. Bohme, Int. Rev. Phys. Chem. **13,** (1994) 163.
2 D. K. Bohme, in *Buckminsterfullerene-ion chemistry in the gas phase*, 1995, p. 1465.
3 S. Petrie, H. J. Becker, V. I. Baranov, and D. K. Bohme, Int. J. Mass Spec. Ion Proc. **145,** (1995) 79.
4 S. Petrie and D. K. Bohme, Can. J. Chem. **72,** (1994) 577.
5 G. Javahery, S. Petrie, H. Wincel, J. Wang, and D. K. Bohme, J. Am. Chem. Soc. **115,** (1993) 5716.
6 A. Hirsch, Q. Li, and F. Wudl, Angew. Chem. Int. Ed. Engl. **30,** (1991) 1309.
7 R. A. Marcus, J. Chem. Phys. **21,** (1952) 359.
8 S. Petrie, G. Javahery, J. Wang, and D. K. Bohme, J. Phys. Chem. **96,** (1992) 6121.
9 J. J. Stry and J. F. Garvey, Int. J. Mass Spec. Ion Proc. **138,** (1994) 241.
10 L. S. Sunderlin, J. A. Paulino, J. Chow, B. Kahr, D. Ben-Amotz, and R. R. Squires, J. Am. Chem. Soc. **113,** (1991) 5489.
11 J. H. Callahan, S. W. McElvany, and M. M. Ross, Int. J. Mass Spectr. Ion Proc. **138,** (1994) 221.
12 J. F. Christian, Z. Wan, and S. L. Anderson, J. Chem. Phys. **99,** (1993) 3468.
13 Z. Wan, J. F. Christian, and S. L. Anderson, Phys. Rev. Lett. **69,** (1992) 1352.
14 Z. Wan, J. F. Christian, and S. L. Anderson, J. Chem. Phys. **96,** (1992) 3344.
15 Z. Wan, J. F. Christian, Y. Basir, and S. L. Anderson, J. Chem. Phys. **99,** (1993) 5858.
16 J. F. Christian, Z. Wan, and S. L. Anderson, J. Phys. Chem. **96,** (1992) 3574.
17 Y. Basir, Z. Wan, J. F. Christian, and S. L. Anderson, Int. J. Mass Spec. Ion Proc. **138,** (1994) 173.

[18] Y. Basir and S. L. Anderson, Chem. Phys. Lett. **243,** (1995) 45.
[19] Y. Cui and L. Liu, Phys. Rev. B **56,** (1997) 3624.
[20] T. Kunert and R. Schmidt, Phys. Rev. Lett. **86,** (2001) 5258.
[21] L. M. Roth, Y. Huang, J. T. Schwedler, C. J. Cassady, D. Ben-Amotz, B. Kahr, and B. S. Freiser, J. Am. Chem. Soc. **113,** (1991) 6298.
[22] Y. Huang and B. S. Freiser, J. Am. Chem. Soc. **113,** (1991) 8186.
[23] G. Javahery, S. Petrie, J. Wang, and D. K. Bohme, Chem Phys. Lett. **195,** (1992) 7.
[24] D. K. Bohme, Chem. Rev. **92,** (1992) 1487.
[25] S. W. McElvany and S. B. H. Bach, A.S.M. Conf. Mass Spectrom. Allied Topics **39,** (1991) 422.
[26] G. Javahery, H. Wincel, S. Petrie, and K. D. Bohme, Chem. Phys. Lett. **204,** (1993) 467.
[27] S. Petrie, J. Wang, and D. K. Bohme, Chem. Phys. Lett. **204,** (1993) 473.
[28] G. Javahery, H. Becker, S. Petrie, P. C. Cheng, H. Schwarz, L. T. Scott, and D. K. Bohme, Org. Mass Spectr. **28,** (1993) 1005.
[29] G. Jahavery, S. Petrie, J. Wang, X. Wang, and D. K. Bohme, Int. J. Mass Spectrom. Ion Processes **125,** (1993) R13.
[30] J. M. Weber, K. Hansen, M. W. Ruf, and H. Hotop, Chem. Phys. **239,** (1998) 271.
[31] B. Brunetti, P. Candori, R. Ferramosche, S. Falcinelli, F. Vecchiocattivi, A. Sassara, and M. Chergui, Chem. Phys. Lett. **294,** (1998) 584.
[32] B. Brunetti, P. Candori, S. Falcinelli, F. Vecchiocattivi, A. Sassara, and M. Chergui, J. Phys. Chem. A **104,** (2000) 5942.
[33] K. Hansen, K. Hoffmann, and E. E. B. Campbell, J. Chem. Phys. **119,** (2003) 2513.
[34] J. M. Blatt and V. F. Weisskopf, *Theoretical Nuclear Physics* (Dover, New York, 1979).
[35] K. Hansen, Philos Mag. B **79,** (1999) 1413.
[36] E. E. B. Campbell, K. Hansen, K. Hoffmann, G. Korn, M. Tchaplyguine, M. Wittmann, and I. V. Hertel, Phys. Rev. Lett. **84,** (2000) 2128.

Chapter 5

COLLISION INDUCED DISSOCIATION

1. SINGLY-CHARGED FULLERENE PROJECTILE IONS

Collision-induced dissociation (CID) is a powerful technique to investigate the structure and stability of molecular ions. The fragmentation of fullerene ions induced in collisions with atomic and molecular targets has been widely studied since macroscopic amounts of fullerenes were first made available. In the first report [1] a reversed-geometry, double-focusing mass spectrometer, with a collision cell situated between the magnetic and electric sectors, was used to study the CID reactions of mass-selected C_{60}^+ and C_{70}^+ at laboratory collision energies in the range 5-8 keV. Fragmentation mass spectra were reported down to C_{40}^+ with mass peaks separated by 24 u (C_2). In the same paper, similar mass spectra for the CID of doubly charged fullerenes and also from charge inversion of negatively charged fullerenes in collisions with O_2 were reported. The mass spectra were very similar to those reported some years earlier by the Smalley group for the photodissociation of mass-selected fullerenes using ns laser pulses [2]. A number of other groups subsequently reported related results using a variety of target gases [3-9]. In these experiments, the laboratory energy was mainly fixed and the centre-of-mass energy was varied by changing the target gas. For sufficiently high excitation energies, a bimodal fragment pattern was observed [3,4,6,7]. The development of such a collisionally induced bimodal fragment pattern in shown in Fig. 5.1(a). These spectra were measured with a time-of-flight mass spectrometer and are for collisions between C_{60}^+ and

Ar atoms [10]. The ion intensities have not been corrected for detection efficiency so that the relative intensity of the small fragment ions is exaggerated in this plot [11]. An extract of the fragment ion spectrum obtained for collisions with Ar at a centre-of-mass collision energy of 100 eV is shown in Fig. 5.1(b). This very clearly shows the cross-over at $n = 32$ from larger fullerene-like fragments with the characteristic C_2 spacing between the mass peaks, to the small, predominantly ring isomers, separated in mass by a carbon atom. The geometry of the various carbon cluster ions has been studied in beautiful gas ion chromatography experiments. The geometry can be determined from a measurement of the drift time of ions through a buffer gas. Fig. 5.2 shows the mobility measured for different carbon cluster ions [12]. The results can be divided into different families depending on the geometry. Fullerenes have a high mobility due to their compact structures, and can be observed down to $n = 30$. Many different isomers can co-exist for cluster sizes of around $n = 30$ but a simple ring structure is the most likely isomer for $10 < n < 20$. Below this, linear structures dominate.

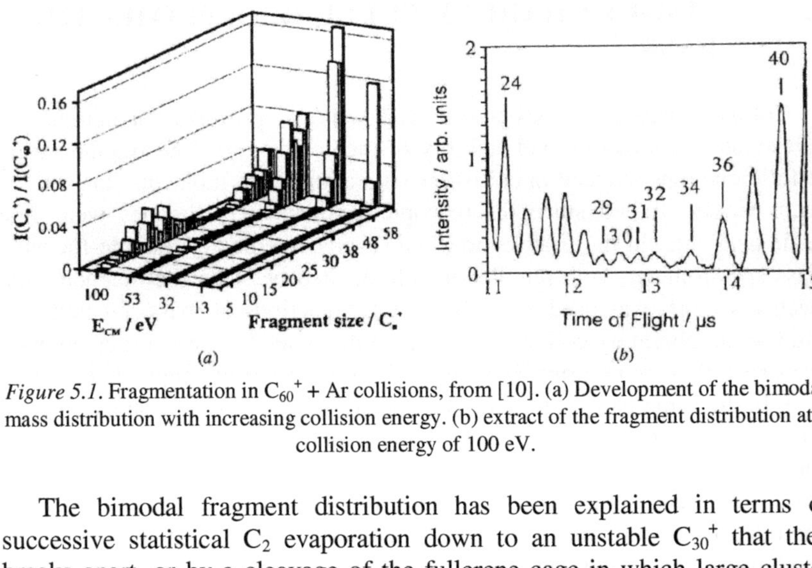

Figure 5.1. Fragmentation in C_{60}^+ + Ar collisions, from [10]. (a) Development of the bimodal mass distribution with increasing collision energy. (b) extract of the fragment distribution at a collision energy of 100 eV.

The bimodal fragment distribution has been explained in terms of successive statistical C_2 evaporation down to an unstable C_{30}^+ that then breaks apart, or by a cleavage of the fullerene cage in which large cluster fragments are emitted. For low excitation energies it is now clear that the dominant fragmentation mechanism is successive evaporation of C_2. As the excitation energy increases there is a small but increasing probability for the emission of larger neutral fragments such as C_4 or C_6. The initial fragmentation process on the picosecond time scale can, however, if the excitation energy is high enough, lead to a catastrophic break-up of the

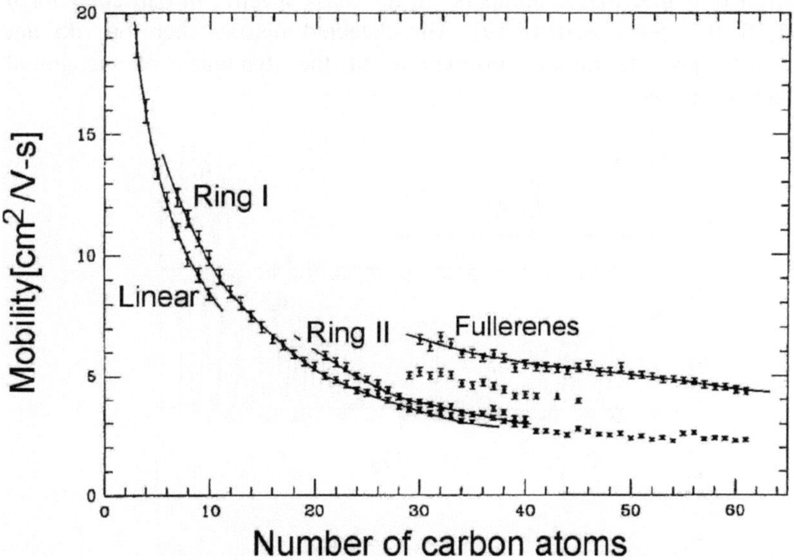

Figure 5.2. Ion mobility versus cluster size for positively charged carbon cluster ions from n = 3 to n = 60. The lines through the data are to guide the eye for the four labeled families of isomers. Adapted from [12].

fullerene cage. Multiple numbers of large fragments result which can undergo a subsequent metastable fragmentation. For fullerene-like fragments this will be C_2 loss whereas for the smaller fragment ions the dominant metastable fragmentation pathway is via the loss of C_3 or a carbon atom. This scenario is supported by some molecular dynamics simulations [13-16] and has been shown explicitly for photodissociation experiments using nanosecond laser pulses [13,17]. Experimentally, the detection of large neutral fragments from collision experiments, under conditions where the positively charged fragments show a bimodal distribution, was reported by McHale et al. [18]. In these experiments, the neutral fragments were post-ionised in subsequent collisions with O_2. A mass distribution closely mirroring that of the positively charged fragments was obtained and provided conclusive evidence for a cleavage mechanism producing large neutral fragments at these collision energies (340 eV in the centre of mass) rather than the dominance of C_2 evaporation, Fig. 5.3. The spectra of neutral fragments, determined in this way, as well as the positive ion fragment spectra from CID of C_{60}^+ are detected on a time scale of microseconds after the initial impact and the ions have time to undergo further metastable cooling evaporations. The detected mass distributions are thus strongly influenced by the thermodynamic stability of the neutral and charged carbon clusters as well as the ionisation potentials. This is the reason for the typical appearance of the "magic numbers" in the mass spectra, in particular for n =7,11,15,19,23,50 [3,4,10,18,19]. The detected masses therefore do not necessarily provide direct information on the dynamics of the initial fragmentation step.

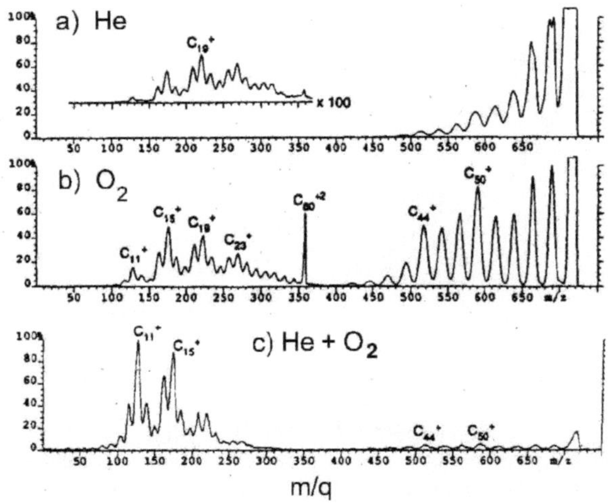

Figure 5.3. a), b) CID mass spectra produced in collisions between C_{60}^+ and He or O_2 collisions at a laboratory collision energy of 8 keV. c) collisional dissociation /reionisation experiment where the fullerene is first dissociated in collisions with He. The charged fragment ions are removed from the beam and the neutral fragments are post-ionised in collisions with O_2. Adapted from [18].

Similar fragmentation patterns are observed for the positively charged fragments produced in collisions with negatively charged ions [1,9,20,21], but the mass distribution of negatively charged fragments looks quite different, Fig. 5.4. Very few negatively charged fullerene fragments are detected. This is attributed to the very high probability of thermal electron emission leading to only neutral fragments that generally go undetected. Thermal electron emission from highly vibrationally excited fullerenes and metal clusters has been addressed in review articles [22,23].

Detailed collision energy dependencies of the fragmentation cross sections in low energy (5-300 eV centre of mass) collisions of positively charged fullerene ions with a variety of targets have been reported [10,24]. The form of the cross sections for production of large fullerene fragments was shown to have a predominant maximum that shifted to larger collision energies as the size of the fragment ions decreased and as the complexity of the target molecules increased. A combination of classical molecular dynamics simulations with statistical RRKM calculations to take into account the metastable fragmentation processes on the microsecond time

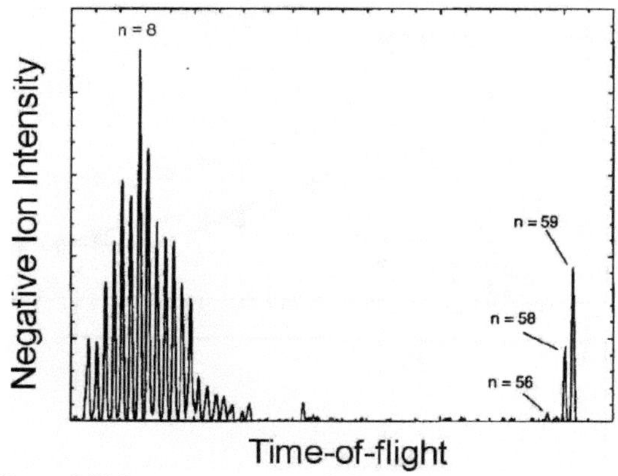

Figure 5.4. Time-of-flight mass spectrum of the negatively charged products of collisions between C_{60}^- and N_2 at a laboratory collision energy of 30 keV. Adapted from [20].

scale of the experiments was able to give reasonable qualitative agreement with the measured cross sections for collisions with Ar [10]. The comparison is shown in Fig. 5.5. The calculations took the total internal excitation energy into account. An average initial internal energy of the projectile ion was estimated (based on the amount of metastable fragmentation of the projectile beam, determined in the absence of collisions) and an estimate of the energy transferred in the collisions was obtained from classical MD simulations. The agreement is quite satisfactory considering that there is a large uncertainty in the absolute value of the experimental cross sections due to uncertainties in determining the target vapour pressure (see Chap. 2.1.3). The relative cross section values as a function of collision energy are much more accurate. Also there are uncertainties associated with the modelling. In particular, the width and form of the initial internal energy distribution in the projectile beam is now known to play a crucial role for determining the appearance of fragmentation spectra, as shown in laser ionisation / fragmentation studies [25]. We do not know the initial internal energy distribution in the fullerene ion beams and can only make a rough estimate of the average internal energy. Also, the parameters used for the RRKM modelling are different from the parameters believed today to provide the best description of the metastable decay of fullerenes (Chap. 3.2.1), however, under the conditions of the laser experiments, this should not make a major difference. The pronounced maxima in the cross sections were seen to be due predominantly to the metastable evaporation of C_2 with the actual shape of the maximum depending on the experimental time window for detection of the fragment ions. The initial internal energy of the projectile

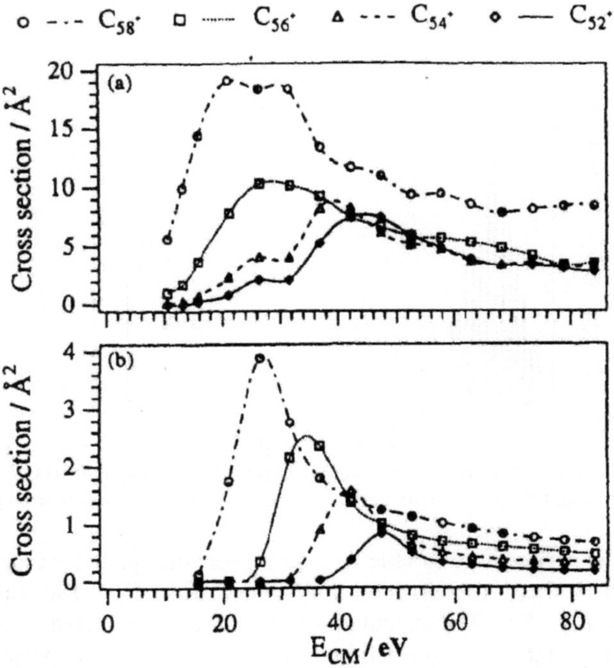

Figure 5.5. Collision energy dependence of the absolute fragmentation cross section in C_{60}^+ + Ar collisions, from [10]. Top: experimental data. Bottom: results from cacluations in which MD simulations are combined with statistical RRKM calculations.

ion does indeed strongly affect the measured form of the cross sections for production of the large fullerene fragment ions. Fig. 5.6 shows a comparison of the total CID cross sections determined for collisions between C_{60}^+ and different targets. Two sets of collisions with Ar are reported where the fullerene projectile ion beam was produced via laser desorption/ionisation with different laser fluence [26]. The cross section labelled Ar(a) was measured using a laser fluence of approximately 20 mJ/cm^2 (as used for the data in Fig. 5.5) whereas Ar(b) was obtained with a laser fluence of approximately 5 mJ/cm^2 which is just slightly above the threshold for production of C_{60}^+. From measurements of black-body radiation from laser desorbed fullerenes [27] it is known that the internal energy of the desorbed molecules scales with the laser fluence. By applying the same simple MD/RRKM modeling to the Ar data of Fig. 5.6 as used to fit the data in Fig. 5.5, it could be estimated that the initial average internal energy of C_{60}^+ was ca. 30 eV in the case of the Ar(a) cross section and ca. 15-20 eV in the case of Ar(b). The lower absolute value of the measured total fragmentation cross section for Ar(b) compared to Ar(a), as well as its shift to higher collision

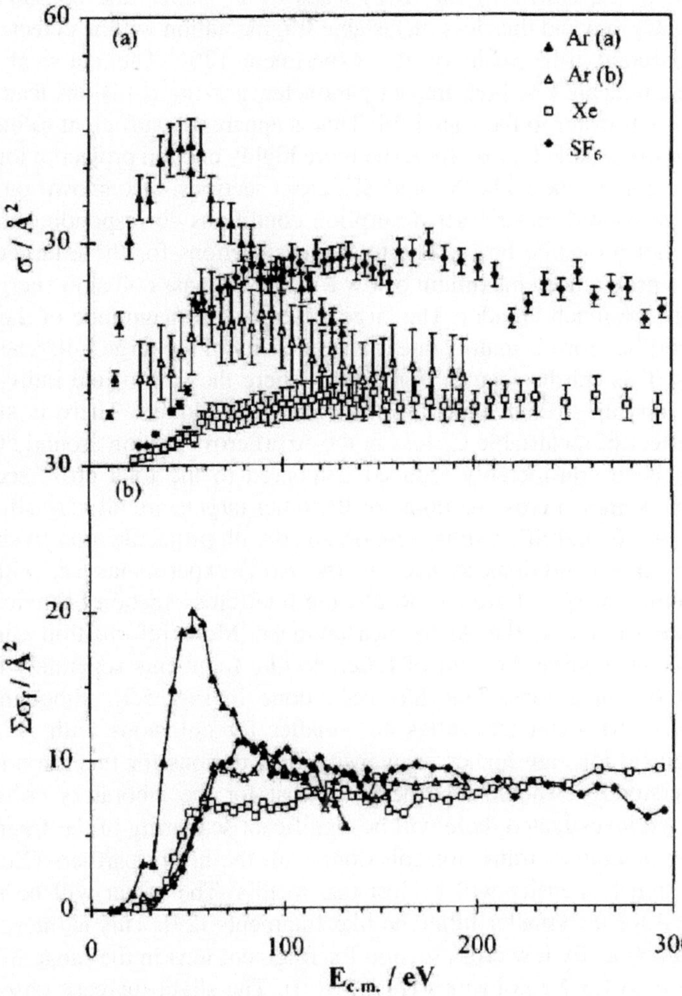

Figure 5.6. (a) Total fragmentation cross sections measured for collisions of C_{60}^+ with Ar, Xe and SF_6 as a function of the centre of mass collision energy. The laser fluence used for production of the fullerene projectile ion beam was 20 mJ/cm^2 for all data except for Ar(b) which was 5 mJ/cm^2. (b) Sum of the individual fragmentation cross sections for C_n^+ production for $n \leq 52$. From [26].

energies, is due to a lower total excitation energy of the C_{60}^+ and the fact that metastable C_2 evaporation which, according to the model, is assumed to be the major contributor to the large peak, can occur on a time scale of more than a few microseconds which was the limit of the experiments. In other words, the lower total excitation energy (initial internal energy plus energy transferred in the collision) for Ar(b) leads to a smaller rate constant for metastable C_2 loss and thus less metastable fragmentation will be detected on the microsecond time scale of the experiment [26]. The classical MD simulations indicate that high impact parameter, grazing collisions lead to a small energy transfer to the cage [28]. This is apparently sufficient to induce substantial metastable C_2 loss from the more highly excited projectile ion but not for the colder ones. The Xe and SF_6 cross sections, also shown on Fig. 5.6, were measured under laser desorption conditions corresponding to the internally hot projectile beam. The total cross sections for these targets do not show a pronounced maximum below a centre of mass collision energy of 100 eV but are much broader. The large difference in magnitude of the SF_6 and Xe cross sections is mainly due to the intensity of the large fullerene-like fragments. This can be seen in Fig. 5.6(b) where the sum of the individual fragment ions up to and including C_{52}^+ has been plotted. There is still a residual effect of metastable C_2 loss in the Ar(a) cross section around 50 eV but it has been considerably reduced compared to the total cross section results. The summed cross sections for the other targets are all very similar. Although the Xe and SF_6 results were obtained with projectile ions produced under the same conditions as used in the Ar(a) experiments i.e. with an initial internal energy of around 30 eV, the total cross section behaviour is much more similar to the Ar(b) measurement. More information can be obtained from plotting the sum of fullerene-like fragments separately from the small fragment ions. This has been done in Fig. 5.7. Although the fullerene-like fragment intensities are smaller for collisions with Xe, the small fragment ions are higher in intensity. The reasons for this are not yet fully understood. It should also be noted that for the laboratory collision energy range investigated there will be significant scattering of the fragment ions in the laboratory frame for collisions with the heavy partners (Xe and SF_6) and signal intensity will be lost due to this. The effect will be most pronounced for the smaller fullerene-like fragments [29]. This is the reason for the exceptionally low cross section for fragment ions in the range $32 \leq n \leq 52$ measured for Xe collisions (Fig. 5.7(a)). The slight increase observed for these cross sections at centre of mass collision energies beyond 200 eV is a consequence of the smaller laboratory scattering angles at these higher energies. It is more difficult to understand the higher cross sections measured for the small fragment ion production. It may be that the stronger interaction potential simply leads to an overall higher collision cross section. Alternatively, it may have to do with the details of the elastic collisions with the carbon cage that lead to an overall higher energy transfer in this collision energy range for the larger Xe atom (see e.g. Fig. 3.3) and therefore a higher tendency for prompt cage break-up to form bimodal fragment distributions

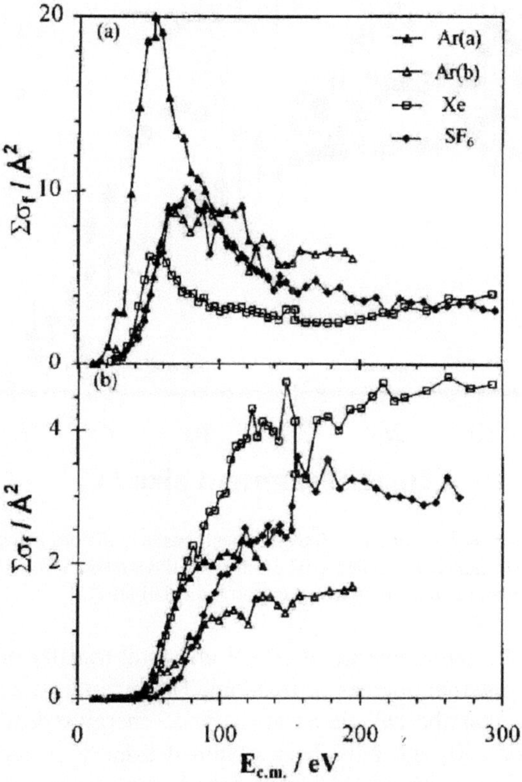

Figure 5.7. Sum of the individual fragmentation cross sections for the data plotted in Fig. 5.6 for (a) $32 \leq n \leq 52$ (b) $n < 26$. From [26].

Some interesting details can be extracted from the cross sections in Fig. 5.7. The fullerene-like fragments from collisions with both Ar(a) and Xe, which were obtained with hot (ca. 30 eV) projectile ions have maxima at the same collision energy. However, the onset for the Xe collisions is slightly shifted to higher collision energies. This was also remarked on in connection with earlier experiments [10] where the appearance energies of the fragments were discussed. These are plotted in Fig. 5.8. The appearance energies measured in collisions with Ar and Ne are very similar over the entire fragment ion mass range. In contrast, the appearance energies in Xe collisions are shifted to higher energies with the SF_6 thresholds lying even higher for loss of more than 10 carbon atoms. The appearance energies for Ar and Ne are in very good agreement with maximal entropy calculations

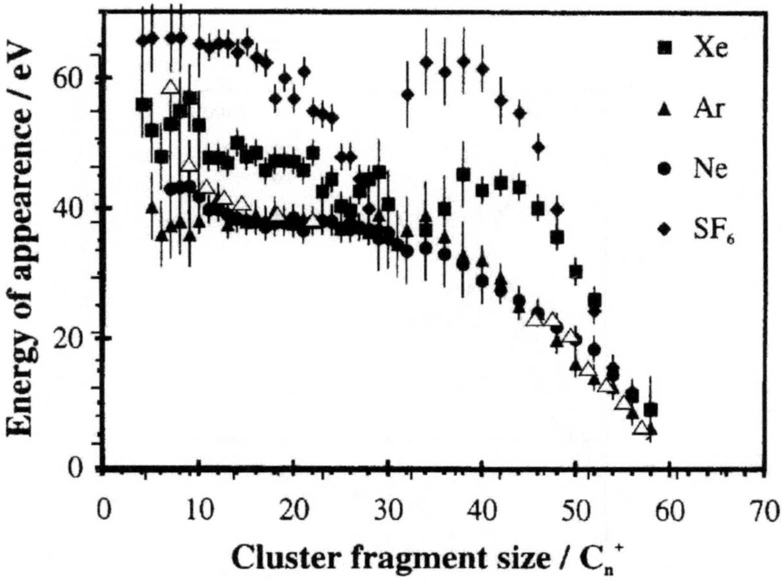

Figure 5.8. Appearance energies of the various fragment ions in collisions between C_{60}^+ and rare gas atoms or SF_6 molecules. Filled symbols are experimental data from [10]. Open triangles are from maximal entropy calculations [25].

assuming an initial internal energy of 30 eV and total transfer of the centre of mass energy to internal degrees of freedom. These results are also shown on Fig. 5.8. Note that the calculated appearance energies differ from the original publication [30] and have been obtained using a corrected energy scale [25]. Originally, with the incorrect energy scale in the calculations, better agreement was obtained between the maximal entropy model and Xe collisions [30]. The lower appearance energies measured for Ar and Ne were therefore attributed to impulsive "knock-out" of fragments from the cage without the transfer of large amounts of energy to vibrational degrees of freedom. Some evidence for such processes is discussed below but we now believe that this is not the main cause of the discrepancy between the Ar, Ne and Xe results in Fig. 5.8. Instead, we relate it to the greater tendency for transient endohedral fullerene formation in the collisions with the smaller rare gas atoms (see Chap. 8). This provides a very efficient mechanism for transfer of the entire centre-of-mass collision energy to internal degrees of freedom of the cage for those collision trajectories that lead to capture. Endohedral fullerene formation is much less efficient for collisions with Xe and is expected to occur only for collision energies higher than most of the fragment appearance energies. Therefore there is a much greater tendency for elastic scattering with cage components leading to transfer of energy less than the centre of mass value. At the centre-of-mass collision energies of interest, the laboratory energy of the fullerene projectile in Xe collisions only covers the range up to slightly less than 2000 eV. For these energies, according to the simple model discussed in Chap. 3, we would expect the Xe to interact with a large region of the cage surface leading to a rather low transfer of energy due to elastic scattering with the cage surface (Fig. 3.3). The even higher appearance energies for collisions with SF_6 can be rationalised by considering the partition of energy between the internal degrees of freedom of the fullerene projectile and the internal degrees of freedom of the molecular target. The drop of the threshold energies at n = 30 for SF_6 is due to the formation of doubly charged fullerene ions in these collisions and the appearance energies plotted in Fig. 5.8 are actually for C_{2n}^{2+} in the region between $n = 20$ and 30.

The onset of the fullerene-like fragments and the position of the total cross section maximum are shifted by at least 10 eV to higher collision energies for Ar(b) compared to Ar(a) (Fig. 5.5a). This corresponds well to the difference in internal energy of the projectile ions in the two series of measurements. Th SF_6 cross section for fullerene-like fragments is very similar to the Ar(b) cross section. This is mainly a consequence of the statistical distribution of excitation energy among all vibrational degrees of freedom in the collision system leading to less vibrational excitation of C_{60}^+ for a given collisional energy loss (remember that the fullerene ions in the collision experiments with Xe and SF_6 have a high initial internal energy of ca. 30 eV). At collision energies around the maxima of the cross sections this energy difference is roughly the same as the difference in internal excitation between the fullerenes used in collisions Ar(a) and Ar(b).

There is a significant difference in the small carbon fragment ion mass distributions measured for Xe and SF_6 collisions. A comparison is shown in Fig. 5.9. The distribution is clearly shifted towards the low mass range for SF_6. There are also additional mass peaks in the SF_6 spectrum, doubly charged fullerene ions and C_nF^+ that are reaction products of the collision [26]. There is no evidence for corresponding reaction products with fullerene-like fragment ions. The enhanced formation of singly-charged small fragments may be explained by the ionisation of small neutral carbon cluster fragments in the course of the collision. Fragments produced on collision induced dissociation of the SF_6 target have high electron affinities e.g. EA (F) = 3.4 eV and EA(SF_5)=3.7 eV [31], much higher than the parent molecules, EA(SF_6) = 1.05 eV. These fragments can strip an electron from the neutral carbon fragments during the course of the collision. The formation of C_nF^+ competes with the charge transfer reaction [26]. This implies that the carbon cage must break-up on a very short time scale (a few hundred fs) in order for the small neutral carbon fragments to react with the fragmentation products from the target. This time scale for cage break-up is confirmed by MD simulations and also by surface collision experiments where the neutral fragments can pick-up an electron from the surface (see Chap. 10).

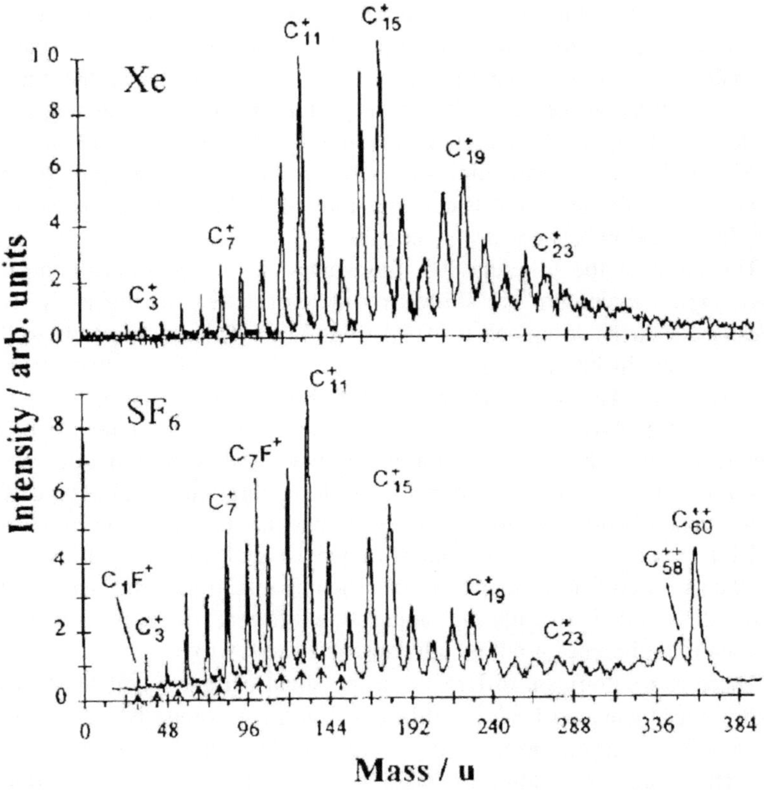

Figure 5.9. Time-of-flight mass spectra of products from collisions between C_{60}^+ and Xe (top) and SF_6 (bottom) at a centre of mass collision energy of 236 eV. Only the small fragment range of the bimodal fragment distribution is shown. The arrows indicate C_nF^+, n = 1-11. From [26].

The results of the low energy collision experiments with fullerene projectile ions can be well explained within the context of statistical models. However, for very small target atoms there is some evidence for an impulsive "knock-out" mechanism. Classical MD simulations have shown trajectories where individual atoms are "knocked out" of the cage in elastic collisions [32]. Experimental evidence for this kind of mechanism has been found for collisions at 50 keV (lab) with He atoms but not for collisions with Ne or Ar targets [6]. The detection of a small signal due to a C_{59}^+ fragment ion in collisions with He, shown in Fig. 5.10, was taken as providing evidence for the impulsive "knock out" of a cage atom without the transfer of large amounts of energy to vibrational degrees of freedom. Non-adiabatic MD simulations also showed trajectories corresponding to such a process but for collisions with fast Ar ions [33]. However, due to the large amount of electronic energy transferred in these collisions, it would not be possible to directly observe the remaining C_{59}^+ since it would very rapidly undergo unimolecular decay to smaller and more stable ions.

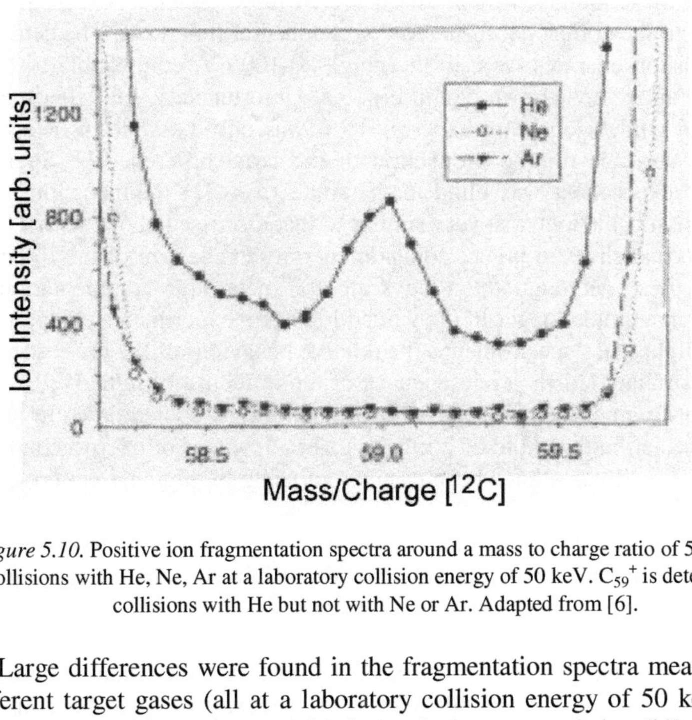

Figure 5.10. Positive ion fragmentation spectra around a mass to charge ratio of 59 for C_{60}^- collisions with He, Ne, Ar at a laboratory collision energy of 50 keV. C_{59}^+ is detected for collisions with He but not with Ne or Ar. Adapted from [6].

Large differences were found in the fragmentation spectra measured for different target gases (all at a laboratory collision energy of 50 keV). The differences could not be explained simply in terms of the different total energy available for excitation (centre of mass energy). The energy transfer to the cage was modeled using the Lindhard model, described in Chap. 3.1.1. The experiments fall into the nuclear stopping regime and the model was used to estimate the energy transferred in each collision as a function of impact parameter. They then considered the probability for a "prompt" fragmentation where single carbon atoms can experience an energy transfer in a single collision that is larger than the energy required for bond breaking all the bonds of the atom in the cage. The rest of the collisionally transferred energy was considered to be statistically distributed among the vibrational degrees of freedom of the remaining cage atoms leading to the normal statistical fragmentation mechanisms discussed above. Using this simple model they were able to obtain very good qualitative agreement with the experimentally measured fragment ion distributions [6].

2. FULLERENE TARGETS

Collision experiments between atomic singly charged ions and neutral fullerenes were carried out using a guided ion beam, apparatus [34-37]. The results of these studies will be discussed in more detail in Chap. 8, with reference to the collisional formation of endohedral fullerenes. The centre of mass collision energies were in the range 20-100 eV, comparable to those discussed above and shown e.g. in Fig. 5.5. Unfortunately, for experimental reasons in guided ion beam experiments it was only possible to determine the cross sections for the production of the large fullerene-like fragment ions. The experiment was blind to the small ($n < 32$) fragment ions. The cross section behaviour was very similar to that discussed above with a clear maximum that shifts to higher collision energies as the size of the fragment ion decreases. Although the time scale for metastable fragmentation os orders of magnitude longer in the guided ion beam experiments compared to time-of-flight mass spectrometry, the relative behaviour of the cross sections is rather similar. This is a consequence of the influence of radiative cooling on the hot fragment ions [38]. The internal energy is cooled down to below the fragmentation threshold on a time scale of a few tens of microseconds.

Many experiments have been carried out using highly charged projectile ions colliding with neutral fullerenes. The charge transfer processes in such collisions are described in Chap. 7. Here we will just focus on the aspects of the experiments and modeling that provide insight into the energy transfer and fragmentation behaviour of the fullerenes.

Experiments involving high energy collisions with singly or multiply charged ions make it possible to scan the impact velocity range where one expects to see a change in the nature of the excitation mechanism. For relatively low energy collisions, energy is predominantly transferred via vibrational excitation (nuclear stopping) while for high energy collisions it is electronic energy transfer (electronic stopping) that is expected to dominate (Chap. 3.1.1). It may also be expected that direct fragmentation processes can be observed for high energy collisions where fragmentation occurs before the energy can be statistically distributed among the vibrational degrees of freedom. Many groups have looked for fingerprints of non-statistical behaviour.

A very spectacular change in fragmentation behaviour as a function of impact velocity was observed by Schlathölter et al. in collisions between He^{q+} projectile beams and neutral C_{60} [39,40]. Mass spectra recorded for collisions with He^+ are shown in Fig. 5.11. At projectile velocities below 0.45 a.u. (corresponding to a collision energy of 20 keV in Fig. 3.6) there is a significant amount of large fullerene fragments indicating the statistical evaporation process seen at lower collision energies and discussed above. These fullerene-like fragments decrease with increasing collision velocity and have almost vanished by 0.45 a.u. while the yield of small fragment ions increases strongly beyond this point. The velocity of 0.45 a.u. is close to where the electronic excitation is predicted to start to dominate over vibrational excitation, according to the Lindhard model. The experiments were nicely explained by invoking the concepts of the Lindhard model but using a more elaborate modeling procedure [40]. Relatively little energy was shown to be transferred in the elastic collisions between the projectile and target atoms at the high collision velocities of these experiments. It is sufficient to produce evaporation of C_2 molecules to form large fullerene-like fragments but is not sufficient to lead to the phase transition and bimodal fragment distribution discussed above. The electronic excitations, due to the electronic stopping of the projectile moving through the electron gas of the fullerene, that dominate for $v > 0.45$ a.u., are more efficient at transferring energy to the fullerene. The simulated projectile energy losses are on the order of 100 eV [40] and thus sufficient after energy coupling to vibrational degrees of freedom has occurred to cause break-up of the cage with emission of ring and chain fragments, leading to the mass distributions shown in Fig. 5.11. The observation of multiply-charged fullerene ions is also in agreement with the occurrence of strong electronic excitations, even for the lowest velocity studied. This interpretation of the experimental results has been qualitatively confirmed by nonadiabatic MD simulations, discussed in Chap. 3.3.3 [33]. In this approach, electronic and vibrational degrees of freedom are treated simultaneously and self-consistently by combining time-dependent density functional theory with classical MD. Snapshots of trajectories for collisions between Ar^+ and C_{60} are shown in Fig. 3.12. The low velocity (0.02 a.u.) trajectory leads to immediate strong deformation of the cage corresponding to the nuclear stopping regime. For the high velocity (0.45 a.u.) trajectory, the ion seems to pass straight through the fullerene cage without causing any significant deformation but transfers a substantial amount of energy to electronic excitation. The electronic excitation relaxes on a time scale of a few hundred femtoseconds [41] thereby exciting vibrational degrees of freedom and ultimately causing fragmentation. There is no reason, from the MD simulations, to invoke exotic fragmentation mechanisms to explain the appearance of the mass spectra. The strongest experimental signature of the crossover between nuclear and electronic stopping should be found in angular distributions of the fragments. Such experiments have not yet been reported.

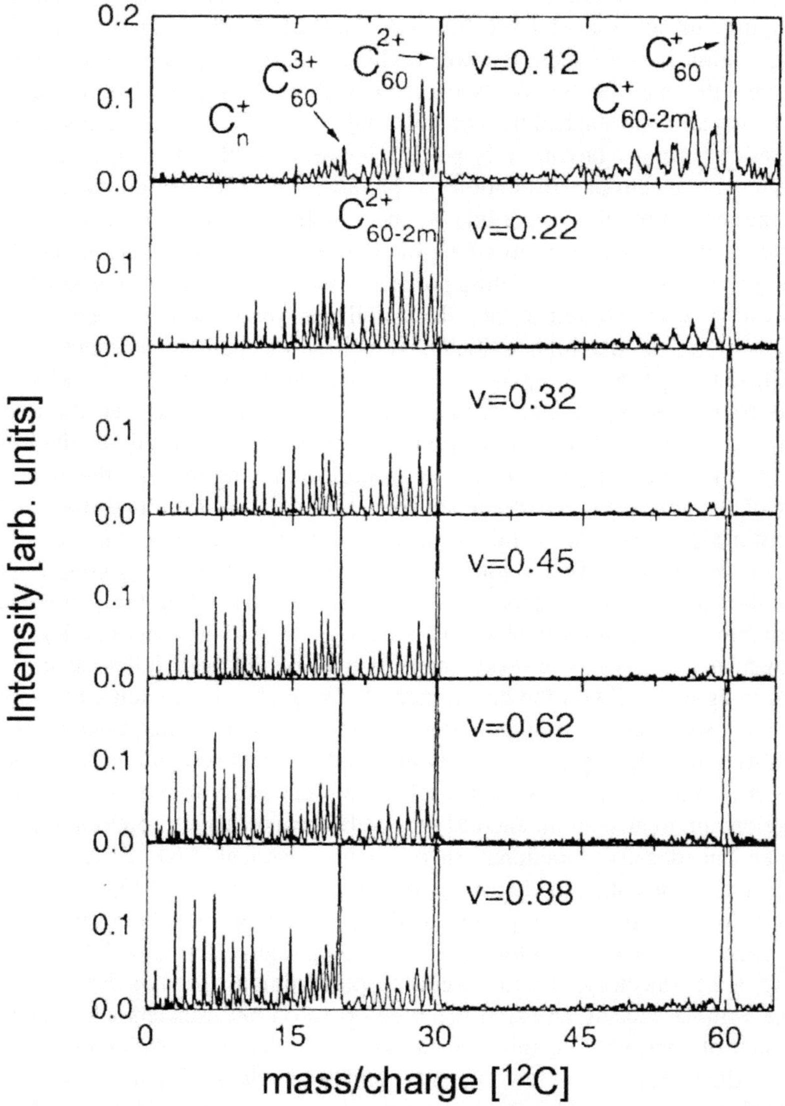

Figure 5.11. Mass spectra obtained for $He^+ + C_{60}$ collisions [39,40]. At slow projectile velocities (< 0.45 a.u.), fragmentation is dominated by statistical evaporation of C_2 units. With increasing projectile velocity the yield of small fragment ions increases while that of the large fullerene-like fragments decreases.

Calculations of the energy transfer versus collision velocity are shown in Fig. 5.12 for three different projectile ions and compared with TRIM

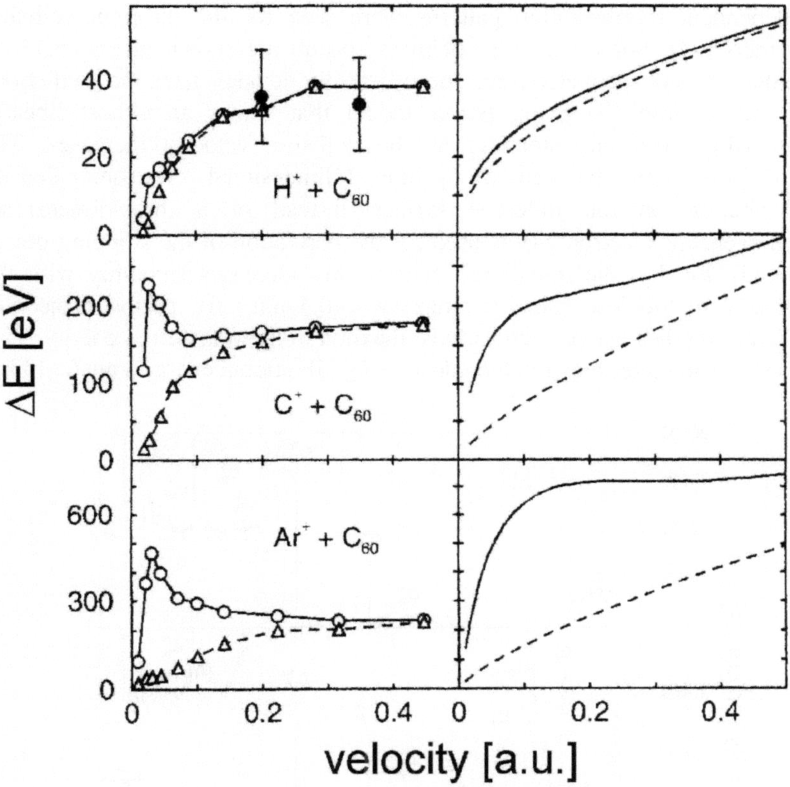

Figure 5.12. left: results of nonadiabatic MD simulations for the total kinetic energy loss (open circles) and its electronic contribution (open triangles) as a function of collision velocity for collisions between different ions (H^+, C^+, Ar^+) with C_{60} and a fixed collision geometry with b = 2 a.u. Experimental data for $H^+ + C_{60}$ are shown by filled circles [42]. Right: TRIM calculations [43] for the same collision systems. Solid lines: total kinetic energy loss. Dashed lines: electronic contributions. Adapted from [33].

calculations [33]. There are also some data points obtained for proton collisions [42] shown on the plot. The total energy transfer and the component due to electronic excitation alone are shown. Some important differences can be seen between the nonadiabatic MD calculations and the TRIM code. There are large differences in the absolute values of the energy transfer for the different ions. This will result in qualitatively different fragment distributions. More surprisingly, the calculations provide velocity-independent excitation energies with dominating electronic excitation in the high velocity range (0.25-0.5 a.u.) for all three ions. These two results provide an explanation for the outcome of collisions with different ions in the high velocity range (v > 0.4 a.u.) where very different but velocity independent fragmentation patterns were seen for the different collision partners [44]. Some examples of mass spectra are shown in Fig. 5.13. A velocity-independent electronic energy transfer deviates from the predictions of the electronic stopping power model that shows an almost linearly increasing electronic stopping as the collision velocity increases. The saturation could be caused by the 2-dimensional electronic density distribution on the fullerene surface instead of a three-dimensional homogeneous electron gas as used for the derivation of the stopping power [43,45]. Finally, the transferred energy only increases smoothly with the velocity in the low velocity range (v < 0.3 a.u.) for proton projectiles whereas the heavier projectiles have maxima in the transferred energy. This results in an increasing and then decreasing vibrational energy transfer [33].

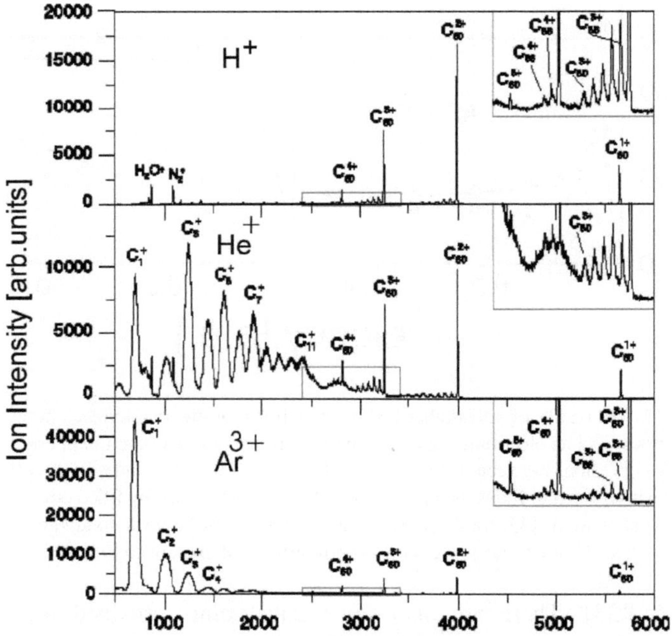

Figure 5.13. Time-of-flight spectra of ions produced in collisions with H^+, He^+ (200 keV laboratory collision energy) and Ar^{3+} (600 keV laboratory collision energy). Adapted from [44].

3. COINCIDENCE EXPERIMENTS

Coincidence experiments have provided more information on the fragmentation dynamics. Often these experiments have involved higher projectile charge states so that it is not always possible to directly compare the results with the low energy, singly-charged fullerenes but they can be compared with the energetic atomic ion - fullerene collisions discussed in 5.2, above. An early study, carried out for collisions between doubly charged C_{60}^{2+} and He at a collision energy of 120 keV showed that C_n^+ fragments as large as $n = 9$ could be measured in coincidence with heavy fullerene-like fragments [46]. Fig. 5.14 shows the coincidence signal obtained in these experiments. Coincidence experiments have also been reported for collisions between neutral fullerenes and highly charged ions [47,48]. In these experiments, the observed fragmentation behaviour was strongly dependent on the charge state of the fullerene after collision as well as the amount of energy transferred to internal degrees of freedom. Experiments in which fragments from 75 keV collisions between C_{60}^- and H_2 were measured in coincidence are particularly revealing [49]. The apparatus used for these studies is shown in Fig. 2.13. The experiments showed that the emission of species larger than C_2 is competitive with C_2 emission in events that leave a large fullerene fragment. The measurements of small fragment ions obtained in coincidence with C_{50}^+ is shown in Fig. 5.15. Small fragments with both odd and even numbers of carbon atoms are found in coincidence with fullerene-like ions, indicating subsequent unimolecular decay of the light fragment if the initial step is a binary fragmentation. The emission of small odd numbered neutral clusters (C_3) is the most energetically favourable unimolecular decay path for the small fragment ions [13]. The sum of the number of carbon atoms of the two species detected in coincidence is generally less than 60 showing that an initial binary fragmentation (or higher) must be accompanied by some neutral fragments. The experimental results were modeled assuming an initial binary fragmentation with subsequent fragmentation estimated according to RRKM theory (similar to [10], discussed above) [50]. The important features and correlations in the fragmentation yields measured in the experiments could be well reproduced assuming an exponential distribution of excitation energies. Such an excitation energy distribution is in agreement with the results of MD simulations for H_2 [51] and He [32] collisions for similar centre of mass collision energies. A different model was put forward to explain the fragmentation pattern observed in collisions between positive fullerene ions and H_2 at 20-200 keV laboratory collision energy [5,52]. here the authors assumed that multiple collisions between the target and carbon atoms in the cage could occur, inducing a large rift in the cage structure. Successive C_2

evaporation from the damaged region could then occur befor the rift "re-closes".

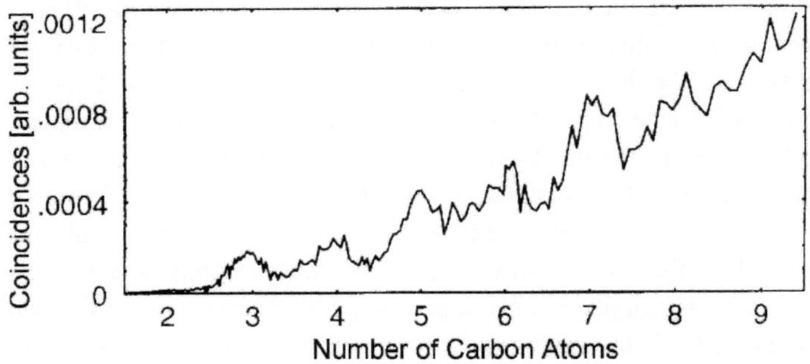

Figure 5.14. Number of coincidences with large fullerene-like clusters registered as a function of cluster size for the products of C_{60}^{2+} + He collisions at a laboratory energy of 120 keV. Adapted from [46].

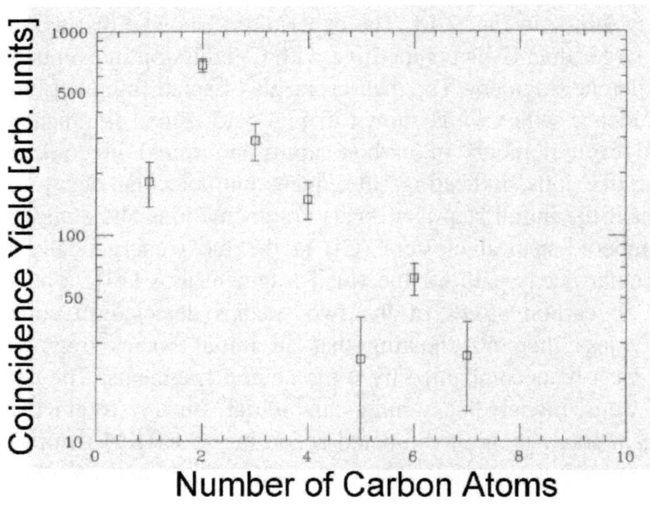

Figure 5.15. Yield of small carbon cluster ions detected in coincidence with C_{50}^{+} for collisions between C_{60}^{-} and H_2 at a laboratory collision energy of 75 keV. From [49].

A multi-coincidence set-up (Fig. 2.14) for investigating the ionisation and fragmentation of C_{60} in collisions with various fast (100-300 keV) low-charged projectile ions provided evidence for both bimodal "fission" and multifragmentation into three or more small charged fragments accompanied by neutral fragments [44]. Examples of time-of-flight fragment spectra in which 1 to 7 particles are detected in coincidence after collisions with Ar^{2+}

are shown in Fig. 5.16 [53]. Each *i*-particle spectrum contains all the detected particles that came from events in which exactly *i* fragments were detected in coincidence. The angular distributions of the charged fragments determined in the same experimental setup were isotropic [53] i.e. no evidence was found for non-statistical "knock-out" or other direct fragmentation processes.

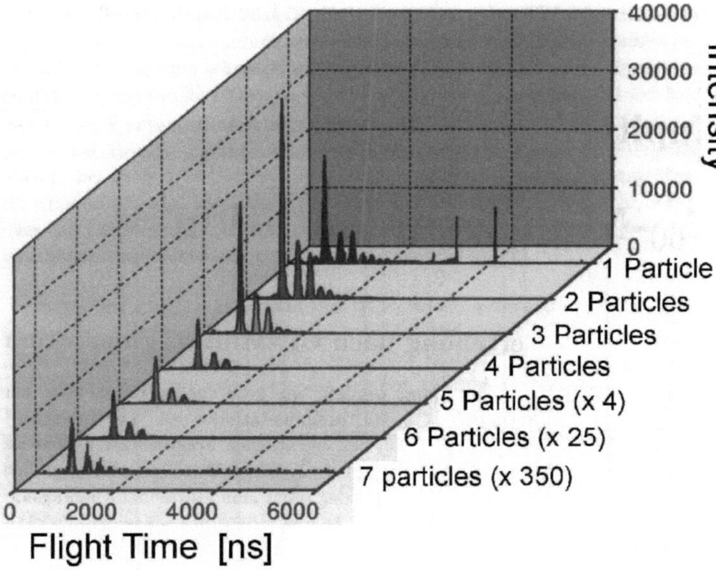

Figure 5.16. Time-of-flight spectra for 1 to 7 coincidentally measured particles after collisions between Ar^{2+} and C_{60} at a laboratory collision energy of 300 eV. Adapted from [53].

Evidence for multi-fragmentation has also been found in measurements of the kinetic energies of fragment ions produced in 2 MeV collisions between Si^{4+} and C_{60} [54].

Detailed modeling of $H^+ + C_{60}$ collisions for the collision energy range of 50-300 keV was carried out using a combination of models but again assuming statistical processes [55]. The ions obtained as the products of such collisions are predominantly fullerene-like fragment ions i.e. the intensity of small fragment ions from a prompt cage break-up is small (5-10%), as can be expected from the calculations of energy transfer in the collisions (Fig. 5.12). At the very high collision velocities considered in these experiments (1-3.5 a.u.), the electronic stopping power actually decreases again with increasing velocity [55]. The coincidence experiments show that these small fragment ions are predominantly from the fission of multiply charged fullerenes [55]. Typical mass spectra for the proton collisions are shown in Fig. 5.17. Three steps were considered in the calculations: firstly the energy transferred to the neutral fullerene target was modeled using the Lindhard model (Chap 3), secondly the electron emission from the excited neutral target was treated using the statistical energy deposition model [56] and thirdly the statistical C_2 emission was considered. A comparison of the experimental and calculated evaporation fraction is shown in Fig. 5.18 as a function of collision velocity. The evaporation fraction is defined as

$$f_e^{q+} = \frac{\sum_{m\geq 1} I\left(C_{60-2m}^{q+}\right)}{I\left(C_{60}^{q+}\right)+\sum_{m\geq 1} I\left(C_{60-2m}^{q+}\right)}. \qquad (5.3.1)$$

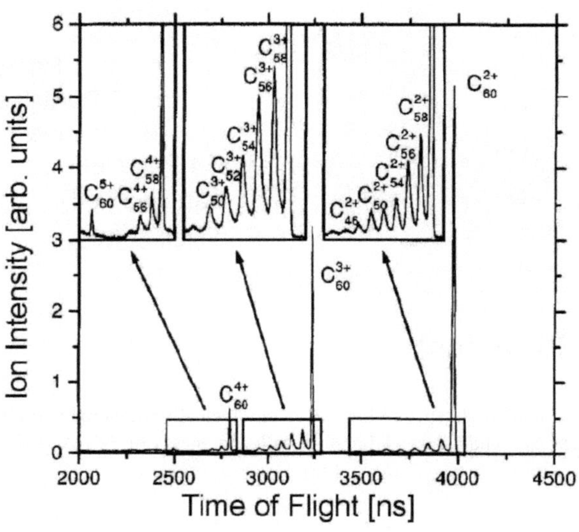

Figure 5.17. Part of the time-of-flight spectra of positively charged products of collisions between protons and C_{60} at a laboratory collision energy of 300 keV. Adapted from [55].

Experimental results obtained by Itoh et al. for even higher collision velocities [57] are shown for comparison on the same plot. The two different experimental data sets fit well together and both sets of results are very well reproduced by the calculations. In order to obtain the agreement with experiment, the parameters used to model the thermal electron emission were adjusted in such as way as to decrease the energy required for electron emission while keeping the exponential pre-factor at the value generally used to model thermionic emission from vibrationally excited fullerenes.

The same effect could be obtained by assuming that fast thermal electron emission occurs from the hot electronic subsystem of C_{60} before the energy is equilibrated with the vibrational degrees of freedom. Such a picture has been used to successfully model the electron emission from fullerenes excited with femtosecond laser pulses or in Penning ionisation experiments (Chap. 4) [41]. The thermal ionisation was found to be an important parameter for correctly describing the relationship between evaporation fraction and fullerene charge state. The higher fragmentation for higher charge states has been observed in many experiments including fs laser excitation and electron impact ionisation as well as charge stripping in collisions with highly charged ions.

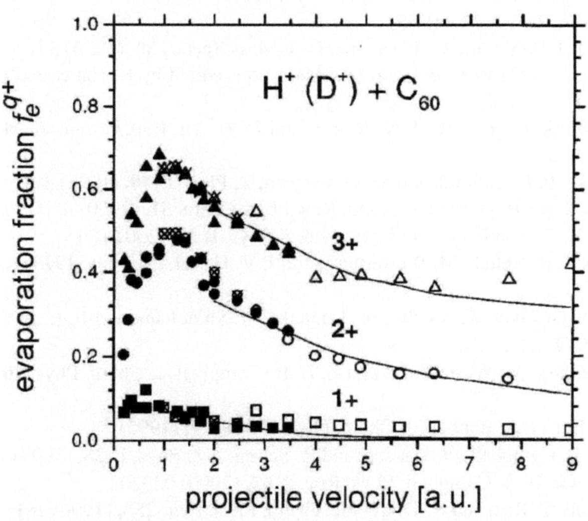

Figure 5.18. Evaporation fractions for charge states q = 1 (squares), q = 2 (circles) and q = 3 (triangles) as a function of the projectile velocity. Solid symbols and crosses are from [55], hollow symbols are from [57]. Lines are the results of model calculations, discussed in the text. Adapted from [55].

REFERENCES

1 A. B. Young, L. M. Cousins, and A. G. Harrison, Rap. Comm in Mass Spec. **5,** (1991) 226.

2 S. C. O'Brien, J. R. Heath, R. F. Curl, and R. E. Smalley, J. Chem. Phys. **88,** (1988) 220.

3 K. A. Caldwell, D. E. Giblin, and M. L. Gross, J. Am. chem. Soc. **114,** (1992) 3743.

4 R. J. Doyle and M. M. Ross, J. Phys. Chem. **95,** (1991) 4954.

5 P. Hvelplund, L. H. Anderson, H. K. Haugen, J. Lindhard, D. C. Lorents, R. Malhotra, and R. Ruoff, Phys. Rev. Lett. **69,** (1992) 1915.

[6] M. C. Larsen, P. Hvelplund, M. O. Larsson, and H. Shen, Eur. Phys. J. D **5,** (1999) 283.
[7] S. W. McElvany and J. H. Callahan, J. Phys. Chem. **95,** (1991) 6186.
[8] T. Weiske, J. Hrusak, D. K. Böhme, and H. Schwarz, Chem. Phys. Lett. **186,** (1991) 459.
[9] D. H. Yu, L. H. Andersen, C. Brink, P. Hvelplund, D. C. Lorents, and R. Ruoff, Mol. Mat. **4,** (1994) 237.
[10] R. Ehlich, M. Westerburg, and E. E. B. Campbell, J. Chem. Phys. **104,** (1996) 1900.
[11] E. E. B. Campbell, G. Ulmer, B. Hasselberger, H.-G. Busmann, and I. V. Hertel, J. Chem. Phys. **93,** (1990) 6900.
[12] G. v. Helden, M.-T. Hsu, P. R. Kemper, and M. T. Bowers, J. Chem. Phys. **95,** (1991) 3835.
[13] H. Hohmann, R. Ehlich, S. Furrer, O. Kittelmann, J. Ringling, and E. E. B. Campbell, Z. Phys. D **33,** (1995) 143.
[14] E. Kim, Y. H. Lee, and J. Y. Lee, Phys. Rev. B **48,** (1993) 18230.
[15] S. G. Kim and D. Tománek, Phys. Rev. Lett. **72,** (1994) 2418.
[16] Y. Xia, Y. Xing, C. Tan, and L. Mai, Phys. Rev. B **52,** (1995) 110.
[17] K. R. Lykke, Phys. Rev. A **52,** (1995).
[18] K. J. McHale, M. J. Polce, and C. Wesdemiotis, J. Mass Spectr. **30,** (1995) 33.
[19] E. E. B. Campbell, in *Clusters of Atoms and Molecules*, edited by H. Haberland (Springer, 1992).
[20] D. Mathur, C. Brink, P. Hvelplund, N. Jensen, and D. H. Yu, Rap. Comm. in Mass Spec. **9,** (1994) 114.
[21] H. Shen, C. Brink, P. Hvelplund, and M. O. Larsson, Z. Phys. D **40,** (1996) 371.
[22] E. E. B. Campbell and R. D. Levine, Annu. Rev. Phys. Chem. **51,** (2000) 65.
[23] J. U. Andersen, E. C. Bonderup, and K. Hansen, J. Phys. B **35,** (2001) R1.
[24] E. E. B. Campbell, R. Ehlich, M. Westerburg, and I. V. Hertel, , Aarhus, 1993 (AIP Press), p. 697.
[25] A. Lassesson, A. Gromov, K. Mehlig, A. Taninaka, H. Shinohara, and E. E. B. Campbell, J. Chem. Phys. **119,** (2003) 5591.
[26] R. Ehlich, H. Sprang, M. Westerburg, and E. E. B. Campbell, J. Chem. Phys. **108,** (1998) 9390.
[27] R. Mitzner and E. E. B. Campbell, J. Chem. Phys. **103 No. 7,** (1995).
[28] R. Ehlich, E. E. B. Campbell, O. Knospe, and R. Schmidt, Z. Phys. D **28,** (1993) 153.
[29] A. V. Glotov and E. E. B. Campbell, Phys. Rev. A **62,** (2000) 033202.
[30] E. E. B. Campbell, T. Raz, and R. D. Levine, Chem. Phys. Lett. **253,** (1996) 261.
[31] S. G. Lias, J. E. Bartmess, J. F. Liebman, J. L. Holmes, R. D. Levin, and W. G. Mallard, J. Phys. Chem. Ref. Data Suppl. **17,** (1988) 872.
[32] R. Ehlich, O. Knospe, and R. Schmidt, J. Phys. B **30,** (1997) 5429.
[33] T. Kunert and R. Schmidt, Phys. Rev. Lett. **86,** (2001) 5258.
[34] J. F. Christian, Z. Wan, and S. L. Anderson, J. Chem. Phys. **99,** (1993) 3468.
[35] Z. Wan, J. F. Christian, and S. L. Anderson, Phys. Rev. Lett. **69,** (1992) 1352.
[36] Z. Wan, J. F. Christian, and S. L. Anderson, J. Chem. Phys. **96,** (1992) 3344.
[37] Z. Wan, J. F. Christian, Y. Basir, and S. L. Anderson, J. Chem. Phys. **99,** (1993) 5858.
[38] K. Hansen and E. E. B. Campbell, J. Chem. Phys. **104 No. 13,** (1996).
[39] T. Schlathölter, O. Hadjar, R. Hoekstra, and R. Morgenstern, Phys. Rev. Lett. **82,** (1999) 73.
[40] T. Schlathölter, O. Hadjar, J. Manske, R. Hoekstra, and R. Morgenstern, Int. J. Mass Spectr. **192,** (1999) 245.
[41] K. Hansen, K. Hoffmann, and E. E. B. Campbell, J. Chem. Phys. **119,** (2003) 2513.
[42] J. Opitz, H. Lebius, S. Tomita, B. A. Huber, P. Moretto-Capelle, D. Bordenave-Montesquieu, A. Bordenave-Montesquieu, A. Reinkoster, U. Werner, H. O. Lutz, A. Niehaus,

M. Benndorf, K. Haghighat, H. T. Schmidt, and H. Cederquist, Phys. Rev. A **62,** (2000) 022705.
[43] J. F. Ziegler and J. P. Biersack, TRIM-code, (1991).
[44] A. Reinköster, U. Werner, and H. O. Lutz, Europhys. Lett. **43,** (1998) 653.
[45] J. Lindhard, V. Nielsen, and M. Scharff, Mat. Fys. Medd. Dan. Vid. Selsk. **36,** (1968) 1.
[46] C. Brink, L. H. Andersen, P. Hvelplund, and D. H. Yu, Z. Phys. D **29,** (1994) 45.
[47] L. Chen, J. Bernard, A. Denis, J. Désesquelles, and S. Martin, Physica Scripta **T80,** (1999) 209.
[48] S. Martin, L. Chen, A. Denis, and J. Desesquelles, Phys. Rev. A **57,** (1998) 4518.
[49] R. Vandenbosch, B. P. Henry, C. Cooper, M. L. Gardel, J. F. Liang, and D. I. Will, Phys. Rev. Lett. **81,** (1998) 1821.
[50] R. Vandenbosch, Phys. Rev. A **59,** (1999) 3584.
[51] G. Seifert and J. Schulte, Physics Letters A **188,** (1994) 365.
[52] D. H. Yu, L. H. Anderson, C. Brink, and P. Hvelplund, Z. Phys. D **29,** (1994) 53.
[53] A. Reinköster, PhD Thesis, (2001) Universtät Bielefeld.
[54] A. Itoh, H. Tsuchida, K. Miyabe, M. Imai, and N. Imanishi, Nucl. Instr. Meth. Phys. Res. B **129,** (1997) 363.
[55] A. Reinköster, U. Werner, N. M. Kabachnik, and H. O. Lutz, Phys. Rev. A. **64,** (2001) 023201.
[56] N. M. Kabachnik, V. N. Kondratyev, Z. Roller-Lutz, and H. O. Lutz, Phys. Rev. A **56,** (1998) 2848.
[57] H. Tsuchida, A. Itoh, Y. Nakai, K. Miyabe, and N. Imanishi, J. Phys. B **32,** (1998) 5289.

Chapter 6

CHARGE TRANSFER IN COLLISIONS WITH HIGHLY CHARGED IONS

One of the presently most active research fields involving fullerene collisions is that of collisions between neutral fullerene targets and highly charged ions. This allows the study of processes in which a large number of electrons is captured from the fullerene targets under the action of the high potential energy of the projectile ions. The studies are fundamentally interesting from at least two main perspectives. One question involves the charge stability of the molecule. How many electrons can one remove from C_{60} and still retain the molecular structure? High impact parameter collisions with highly charged ions provide a means of gently removing electrons from the molecule without transferring significant amounts of energy to vibrational degrees of freedom. This allows the production of high fullerene charge states and their stability can be studied using time-of-flight mass spectrometry or other techniques.

The second questions concerns what happens to the projectile ion. Collisions between highly charged ions and surfaces have been extensively studied for a number of years [1]. So-called "hollow atoms" are produced during such collisions. Hollow atoms are short-lived, multiply excited atoms which carry a large proportion of their electrons in high n levels while inner shells remain transiently empty. This unusual population inversion arises for a time on the order of a few 10's of femtoseconds during the interaction of a slow highly charged ion with a solid surface. Despite the limited lifetime it is possible to obtain information concerning the formation and decay of the hollow atoms by studying electron and X-ray emission as well as the final charge and energy of the scattered projectiles [1]. Fig. 6.1 shows a schematic picture of the formation and decay of hollow atoms at surfaces. In many ways a fullerene behaves like a small piece of bulk matter. Very similar phenomena can occur in collisions between slow highly charged ions and fullerene molecules. The big advantage of these systems is that they are gas phase collisions and the projectile is not immediately destroyed after formation. This makes it possible to probe the projectile properties after the collisions and to study the detailed dynamics with the help of coincidence experiments.

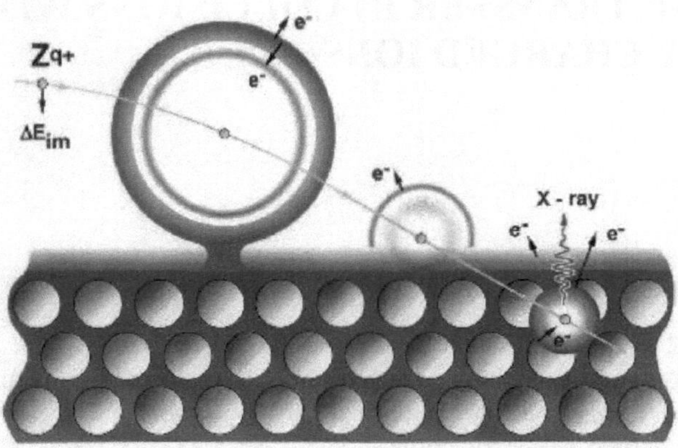

Figure 6.1. Schematic picture of the processes that occur when a slow highly charged ion collides with a surface.

1. HIGH CHARGE STATES OF C_{60}

Multiply charged fullerenes can decay via fission processes due to Coulomb instability in which they emit charged fragments. The Rayleigh limit against fission for charged objects is defined via the equality between the dissociative Coulomb force and the attractive cohesive force [2]. It is difficult to probe this limit for finite highly excited systems such as clusters and large molecules due to their internal excitations that also induce instability. The limit was demonstrated experimentally for the first time only very recently in studies of fission of charged microdroplets [3]. There have been a number of theoretical investigations of the charge stability of C_{60}. The predicted theoretical limit for C_{60}^{q+} ranges from $q = 10$ [4] to $q = 13$ [5] and $q = 16$ [6].

Highly charged ion collisions have been used to produce highly charged C_{60}^{q+} ions from neutral C_{60}. Since electron capture occurs at large impact parameters where only small amounts of energy are transferred to the internal degrees of freedom of the molecule, this is thought to be one of the most promising methods to probe the charge stability limit. Charge states up to $q = 9$ [7] and $q = 10$ [8] have been observed in collisions with Bi^{44+} and Xe^{25+}, respectively. The ionisation and fragmentation patterns obtained in highly charged ion collisions are very similar to those obtained in femtosecond laser ionisation experiments or electron impact experiments. This is not too surprising since the interaction times are similar in all three cases and the interaction predominantly induces electronic excitation. Fig. 6.2 shows a comparison between mass spectra produced by collisions with Ar^{8+} and by 800 nm, 25 fs laser irradiation. By making use of coincidence collision experiments, it is possible to select out events that minimise the transfer of energy to internal degrees of freedom of the fullerene and thus reduce the amount of fragmentation observed in the mass spectra. An example of this is shown in Fig. 6.3 for collisions between Ar^{8+} and C_{60} at a collision energy of 16 keV [9]. Here, the spectra have been recorded in coincidence with different outgoing projectile charge states $Ar^{(8-s)+}$. In Fig. 6.4, we show the evidence for production of $C_{60}{}^{10+}$ in collisions with Xe^{25+} at a collision energy of 280 keV [8].

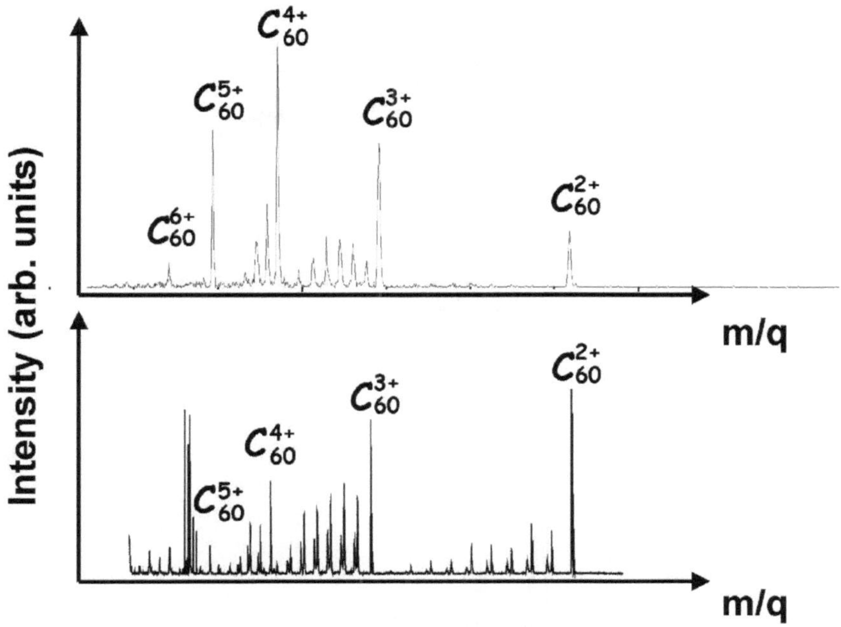

Figure 6.2. Comparison between ionisation /fragmentation spectra for highly charged ion collisions (upper spectrum, collisions with Ar^{8+} at 26 keV [10]) and fs laser excitation (lower for 25fs 800 nm pulses with an intensity of 10^{15} W/cm^2). The large series of peaks in the laser spectrum just below $C_{60}{}^{5+}$ is due to Xe^+, used to calibrate the laser intensity.

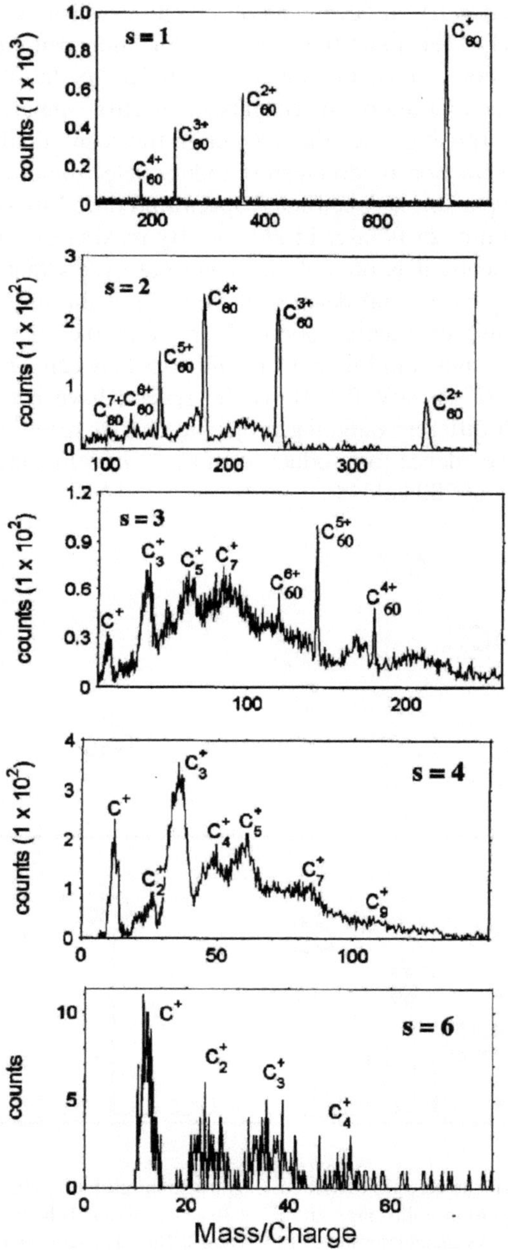

Figure 6.3. Fragmentation spectra measured in coincidence with different outgoing projectile charge states (8-*s*) with *s* = 1-4, 6 for collisions between Ar^{8+} and C_{60} at a laboratory collision energy of 16 keV. Adapted from [9].

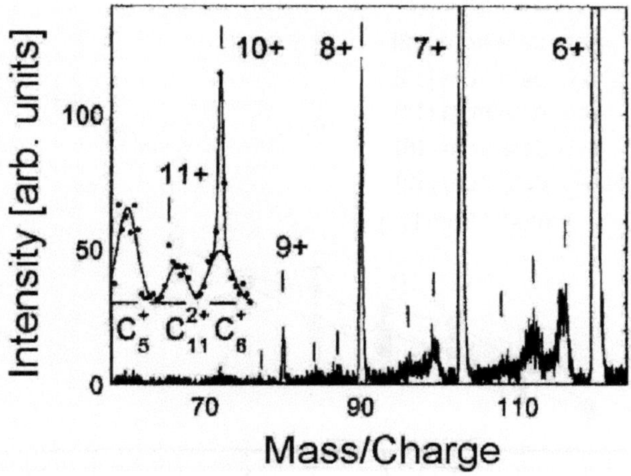

Figure 6.4. Single stop time-of-flight mass spectrum of highly charged C_{60}^{q+} produced in collisions with Xe^{25+} at a laboratory collision energy of 280 keV. The spectrum provides evidence for C_{60}^{10+}, superimposed on a singly charged C_6^+ fragment. Adapted from [8].

Recently, unambiguous evidence has been obtained for C_{60}^{12+} via multiphoton absorption of infrared photons (1800 nm) in femtosecond laser experiments [11]. This is the highest fullerene charge state observed so far.

All the experiments reported so far have used an effusive C_{60} beam source and thus the molecules have a relatively high initial internal vibrational energy (see Chap. 2.1.3). It may be possible to observe even higher charge states if the fullerene targets are initially in the vibrational ground state. Experiments with vibrationally cold C_{60} are possible by using a cluster aggregation source [12] and collision experiments with such a source are expected to become available in the near future [13].

Recent measurements of the kinetic energy release in fragmentation reactions of highly charged C_{60}^{q+} have provided information on the stability of the multiply charged fullerene ions [14]. These experiments were carried out using the coincidence set up shown in Fig. 2.15. The kinetic energy release was determined from the flight time and lateral spread of the fragment ion signals detected with a position sensitive detector. The fragment ions were detected in coincidence with a final projectile charge state. The results of the kinetic energy release measurements for the process where C_{60}^{q+} emits a single C_2^+ fragment are shown in Fig. 6.5. The competing channel where the fullerene emits a neutral C_2 fragment shows a significantly lower value for the kinetic energy release [14]. The measurements are compared with measurements using different methods such as the MIKE (mass-analysed ion kinetic energies, Fig. 2.16) [15,16]

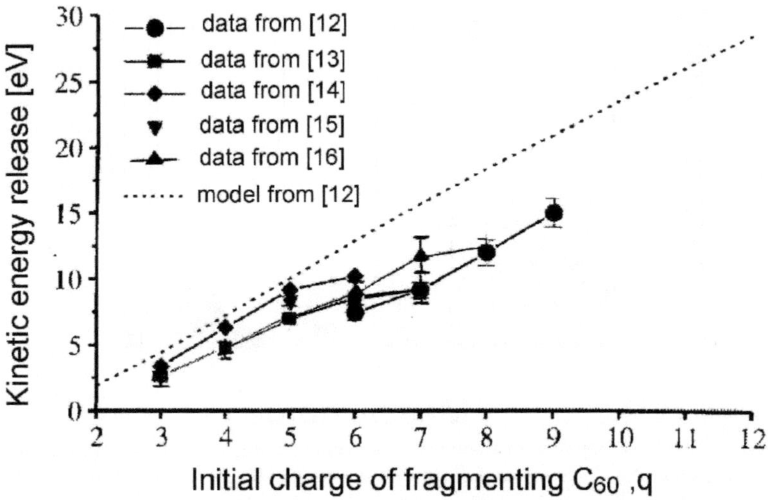

Figure 6.5. Kinetic energy release for the fission process where a multiply charged C_{60}^{q+} ion emits a C_2^+ fragment. Adapted from [14].

and time-of-flight techniques [17,18]. In the latter experiment, the coincidences between two fragment ions from the same event are determined. A simple model was used to interpret and discuss the kinetic energy release measurement results [14]. The two separating fragments are treated as conducting spheres The model was first developed by Näher et al. in 1997 [19]. This was later extended to take the effect of electron transfer during fragmentation into account [20]. This is the basis of the autocharge transfer mechanism [15] in which it is assumed that the fragmentation starts out as a separation between a charged heavy fragment (C_{58}^{q+}) and a light neutral fragment (C_2) which at a rather large critical distance R_C loses an electron to the larger charged fragment. If one considers the potential energies in such a model, Fig. 6.6, one can see that it is energetically more favourable to emit a neutral C_2 than a C_2^+ from C_{60}^{2+}. For $q = 3$, the two potential energy curves cross at a distance of approximately 14 a_0 which is close to but just outside the critical distance for electron transfer, 12.7 a_0. This implies that although the potential energies of the sphere-sphere system are the same for asymmetric fission and evaporation at the crossing point, the potential barrier which the electron experiences as it attempts to move from the lighter to the heavier sphere is higher than its total energy. Under these circumstances the autocharge transfer process would require tunneling through a barrier and is not expected to be efficient. This is in agreement with the presented experimental results [14]. For $q = 4, 5$ the fission potential energy curve lies below the evaporation curve for all distances and therefore

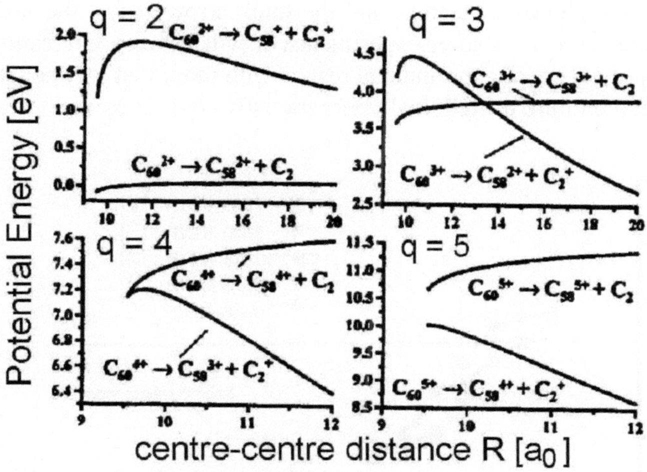

Figure 6.6. The interaction energies for two conducting spheres, used to model the $C_{58}^{(q-1)+}$ + C_2^+ and $C_{58}^{q+} + C_2$ potentials as a function of R for q = 2,3,4 and 5. The potential energies for fission approach zero as R approaches infinity. Adapted from [14].

there is no crossing and direct C_2^+ emission is expected to dominate. The kinetic energy release values estimated from the sphere-sphere model are shown on Fig. 6.5 as the dotted line. It is assumed in the model that the fission products are in their electronic and vibrational ground states. This is not the case in the experiments and the final states will be at least vibrationally excited, both due to the initial temperature of the neutral fullerene target and also to any energy transferred during the collision, and may also be electronically excited. This would yield a smaller difference in relation to the maximum of the potential energy barrier and therefore smaller values of the kinetic energy release.

The kinetic energy release in asymmetric fission is related to the fission barrier, $B_{fiss}{}^q$, via the expression [14]

$$B^q_{fiss} = E_{KER} + D^q + IP(C_2) - IP(C_{58}) \qquad (6.1.1)$$

where D^q is the dissociation energy for C_2 emission from $C_{60}{}^{q+}$. This is plotted as a function the initial charge of C_{60} in Fig. 6.7, using the presently accepted value for the C_2 dissociation energy (10 eV, with the additional assumption that this is q independent) (Chap. 3.2.1) and ionisation potentials estimated in [21]. The sphere-sphere model gives a stability limit for the charge on the fullerene cage of q = 18. The extrapolation of the experimental data (which unlike the model does not involve ions with zero internal energy) gives a critical charge of 11. The theoretical predictions of the stability limit are shown by arrows and the thick arrow shows the highest charge state observed in fs laser experiments. It will be very interesting to compare the present set of experimental results with those that will hopefully be obtained in the future using initially vibrationally cold fullerenes.

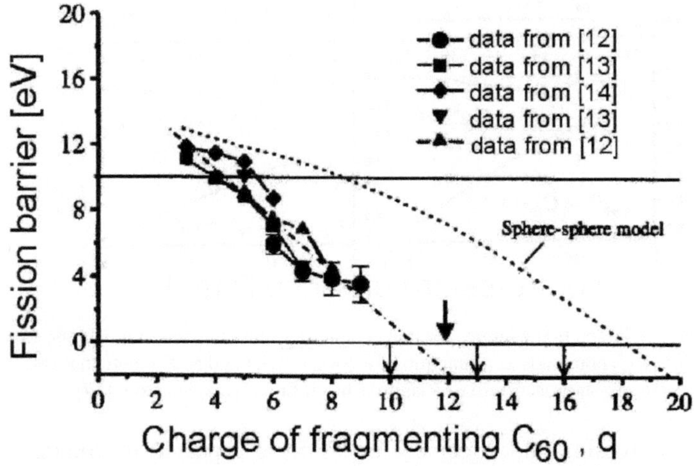

Figure 6.7. Semi-empirical and model fission barriers for $C_{60}^{q+} \rightarrow C_{58}^{(q-1)+} + C_2^+$. The curves show a fit to the experimental data and the results of the sphere-sphere model. The thin arrows indicate theoretical predictions of the stability limit [4-6]. The thick arrow indicates the q=12 charge state observed in laser experiments [11]. Adapted from [14].

The branching ratios between singly-charged light fragments and singly and doubly charged light fragments emitted in asymmetric fission reactions have been studied by Martin et al. in a similar coincidence setup involving collisions between Xe^{25+} and C_{60} [21]. The branching ratios for the number of carbon atoms in a singly-charged light fragment are plotted in Fig. 6.8 for different initial charges, q on the parent molecule. The light fragment obviously shift to smaller numbers of carbon atoms as the parent molecule charge state increases. The experimental results are well reproduced with a simple statistical model showing that the asymmetrical fission of C_{60}^{q+} for q up to 9 can be described by a thermally activated evaporative process. The branching ratios of the total singly- and doubly-charged light fragment intensities are plotted in Fig. 6.9 as a function of charge on the parent molecule. The total branching ratio of these channels increases from less than 2% to 25% as the charge on the parent molecule increases from q = 5 to q = 9. This is because the fission barrier for the emission of the doubly charged light fragment decreases faster than that of the singly-charged emission channel as the parent molecule charge increases [21].

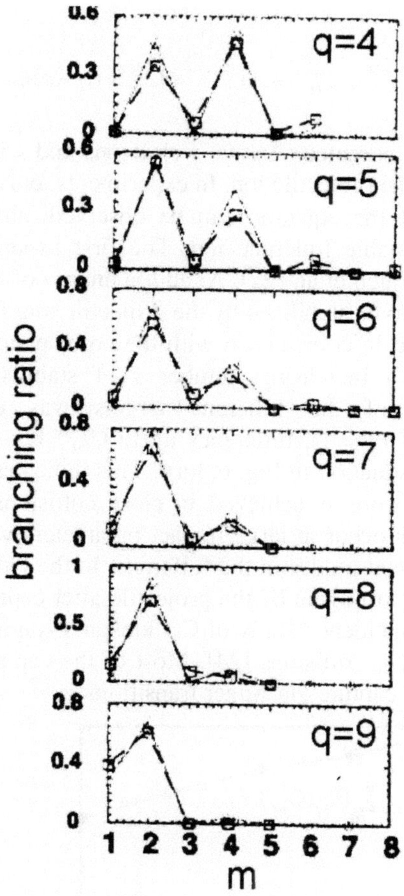

Figure 6.8. Comparison of experimental branching ratios (squares) with a statistical model (circles) for the fission channels $C_{60}^{q+} \rightarrow C_{60-m}^{(q-1)+} + C_m^+$ with q ranging from 4 to 9. Adapted from [21].

2. CHARGE TRANSFER MECHANISMS AND FINAL PROJECTILE STATES

Although many electrons can be removed from the C_{60} in collisions with highly charged ions, not all of the transferred electrons will be stabilised on the projectile ion. A collision system that has been applied by a number of groups to the study of this effect is $Ar^{8+} + C_{60}$ [9,22,23]. The reaction equation therefore reads

$$Ar^{8+} + C_{60} \rightarrow Ar^{(8-q)+} + C_{60}^{q+} \rightarrow Ar^{(8-s)+} + C_{60}^{q+} + (q-s)e^{-} \qquad (6.2.1)$$

where q is the number of extracted ("active") electrons and s is the number of electrons stabilised on the projectile ion. In experiments, only the products on the right-hand side of the equation can be observed, along with any fragments from the metastable fullerene ion. The first experiment of this kind was carried out by Walch et al. [22]. At an ion energy of 80 keV, up to eight electrons were seen to be stabilised by the projectile ion. Projectile ions with fixed s were detected in coincidence with the corresponding fullerene ions and fragments. With increasing number s of stabilised electrons, enhanced fragmentation of the fullerene targets was observed. In coincidence with the Ar^{7+} (s = 1), fullerenes up to C_{60}^{6+} but no fragments were seen (similar to the situation in Fig. 6.3(a)). This indicates that a larger number of stabilised electrons is achieved in close collisions, while pure electron transfer processes occur at large impact parameters with relatively little excitation of vibrational modes in the fullerene. In the latter case, most of the active electrons are lost again by the projectile after capture. This was also concluded from a coincidence study of C_{60} ions and fragments with Ar K x-rays from $Ar^{17/18+}$ + C_{60} collisions [24]. Most of the captured electrons are lost again shortly after capture via Auger transitions.

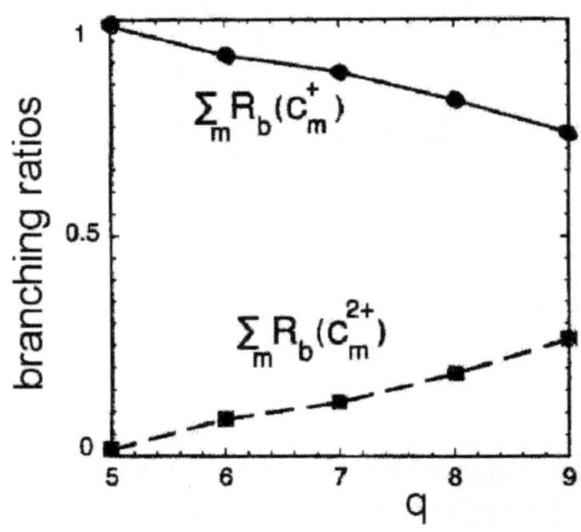

Figure 6.9. Total branching ratios of the singly and doubly charged light fragment emission channels versus the charge state q of the parent ion. Adapted from [21].

Total absolute cross sections σ_s were measured for the formation of stabilised projectile ion states [9,22,23] and are shown in the lower part of Fig. 6.10. Below s = 4 the cross section increases strongly. There is a small maximum around s = 5,6 and then it drops down towards s = 8 (corresponding to projectile neutralisation). From the cross section data in the figure in combination with time-of-flight spectra of the fullerene ions and fragments as a function of s one can obtain absolute cross sections for the production of C_{60}^{q+}, σ_q. Experimental capture radii are then obtained from

$$\sigma_q = \pi\left(R_q - R_{q+1}\right)^2 \qquad (6.2.2)$$

with the total cross section

$$\sigma = \pi R_1^2 = \sum_q \sigma_q . \qquad (6.2.3)$$

R_q decreases from 23 a_0 for $q = 1$ to 14 a_0 for $q = 6$. The values of the capture radii can be understood in terms of an over the barrier model (see Chap. 3.4.1). The active electron is assumed to be transferred to the projectile at the distance where the potential energy of the electron matches its Stark-shifted

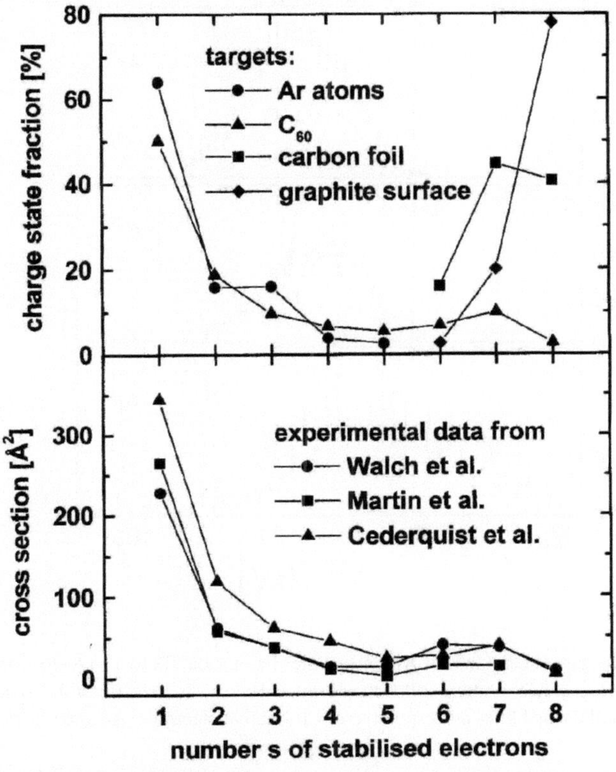

Figure 6.10. Top: charge state fraction $f_8^s = \sigma_s/\Sigma\sigma_s$ of the Ar projectile ions after collision with various targets. Bottom: absolute cross section for stabilisation of s electrons on the projectile after $Ar^{8+} + C_{60}$ collisions. Data from three independent measurements are shown [9,22,23].

binding energy in the fullerene (ion). Similar approaches in more or less sophisticated forms were employed by a number of groups for the analysis of electron transfer in ion-fullerene collisions for various collision systems [9,25-31]. It is interesting to note that the capture radii and cross sections obtained from an over the barrier model in which the positive charges on the C_{60} are assumed to be point-like and to move freely on the fullerene surface are not too different from the results obtained by treating the fullerene as a conducting sphere.

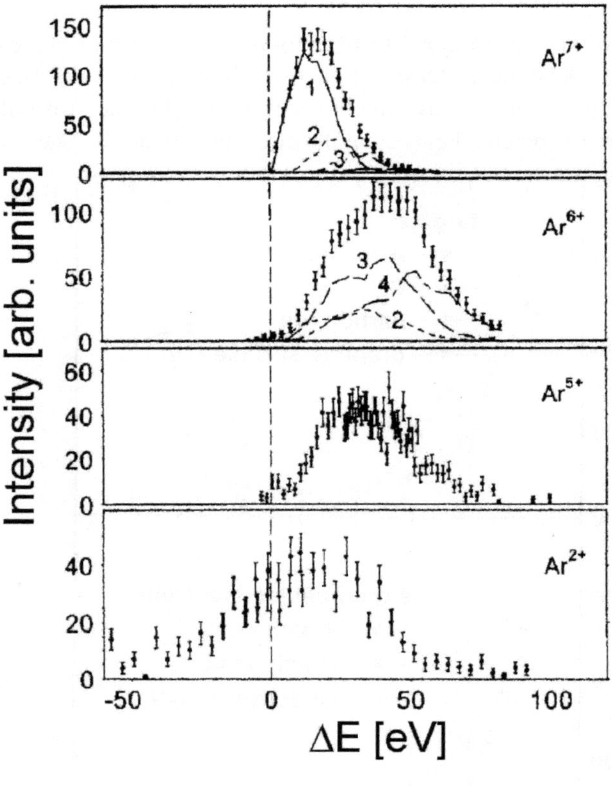

Figure 6.11. Energy gain and loss of Ar^{8+} projectiles having stabilised 1 to 6 electrons after collisions with C_{60} at a laboratory collision energy of 16 keV. In the case of Ar^{7+} and Ar^{6+} ions the full and dashed lines correspond to q = 1-4 active electrons. Adapted from [32].

The relation between Ar^{8+} + C_{60} collisions and collisions of highly charged ions with atoms and surfaces as well as penetration of highly charged ions through carbon foils is shown in the upper part of Fig. 6.10. Here, the projectile charge fraction $f_8^s = \sigma_s / \sum \sigma_s$ is plotted as a function of the number of stabilised electrons, s [9]. For low s the behaviour of collisions with C_{60} is very similar to those with Ar atoms. For $s > 5$, f_8^s increases, similar to observations in surface scattering and passage of the projectile through a carbon foil. The hump around $s = 6$ is present in the C_{60} as well as in the carbon foil data. It is interpreted in terms of equilibration of the charge state on the projectile in small impact parameter collisions, which are frequent in the carbon foil experiments. In the case of C_{60} it is assumed that trajectories through the fullerene cage have a similar effect and are therefore responsible for the hump.

Measurements of the projectile kinetic energy gain in the case of single electron capture allows the assignment of the atomic ion state into which the electron is captured [33]. Since no projectiles with kinetic energy loss or negligible energy gain are observed, it is concluded that no electronic excitations in the C_{60} target are involved in the single electron transfer events (a reason for the fragmentation free spectra, Fig. 6.3). The situation changes when a larger number of electrons is captured from C_{60}. For $s = 6$, the projectile starts to lose kinetic energy as a consequence of strong electronic excitations in C_{60} which are produced in small impact parameter collisions, Fig. 6.11.

REFERENCES

1 H. Winter and F. Aumayr, J. Phys. B **32,** (1999) R39.

2 R. Lord, Philos. Mag. **14,** (1882) 184.

3 D. Duft, H. Lebius, B. A. Huber, C. Guet, and T. Leisner, Phys. Rev. Lett. **89,** (2002) 084503.

4 J. Cioslowski, S. Patchkovskii, and W. Thiel, Chem. Phys. Lett. **248,** (1996) 116.

5 T. Bastug, P. Kurpick, J. Meyer, W. D. Sepp, B. Fricke, and A. Rosen, Phys. Rev. B **55,** (1997) 5015.

6 G. Seifert, R. Gutierrez, and R. Schmidt, Phys. Lett. A **211,** (1996) 357.

7 J. Jin, H. Khemliche, and M. H. Prior, Phys. Rev. A **53,** (1996) 615.

8 A. Brenac, F. Chandezon, H. Lebius, A. Pesnelle, S. Tomita, and B. A. Huber, Physica Scripta T **80B,** (1999) 195.

9 H. Cederquist, A. Fardi, K. Haghighat, A. Langereis, H. T. Schmidt, S. H. Schwartz, J. C. Levin, I. A. Sellin, H. Lebius, B. Huber, M. O. Larsson, and P. Hvelplund, Phys. Rev. A **61,** (2000) 022712.

10 H. Cederquist, , (2001) private communication.

11 V. R. Bhardwaj, P. B. Corkum, and D. M. Rayner, Phys. Rev. Lett., (2003) in press.

12 K. Hansen, R. Mueller, P. Brockhaus, E. E. B. Campbell, and I. V. Hertel, Z. Phys. D **42,** (1997) 153.

13 T. Bergen, X. Biquard, A. Brenac, F. Chandezon, B. A. Huber, D. Jalabert, H. Lebius, M. Maurel, E. Monnand, J. Opitz, A. Pesnelle, B. Pras, C. Ristori, and J. C. Rocco, Rev. Sci. Instrum. **70** 3244.

[14] H. Cederquist, J. Jensen, H. T. Schmidt, H. Zettergren, S. Tomita, B. A. Huber, and B. Manil, Phys. Rev. A **67,** (2003) 062719.
[15] P. Scheier, B. Dünser, and T. D. Märk, Phys. Rev. Lett. **74,** (1995) 3368.
[16] G. Senn, T. D. Märk, and P. Scheier, J. Chem. Phys. **108,** (1998) 990.
[17] L. Chen, J. Bernard, G. Berry, R. Bredy, J. Desesquelles, and S. Martin, Phys. Scr. T **92,** (2001) 138.
[18] S. Tomita, H. Lebius, A. Brenac, F. Chandezon, and B. A. Huber, Phys. Rev. A **67,** (2003) 063204.
[19] U. Näher, S. Bjoernholm, S. Frauendorf, S. Garcias, and C. Guet, Phys. Rep. **285,** (1997) 245.
[20] H. Zettergren, H. T. Schmidt, H. Cederquist, J. Jensen, S. Tomita, P. Hvelplund, H. Lebius, and B. A. Huber, Phys. Rev. A **66,** (2002) 032710.
[21] S. Martin, L. Chen, R. Bredy, J. Bernard, M. C. Buchet-Poulizac, A. Allouche, and J. Desesquelles, Phys. Rev. A **66,** (2002) 063201.
[22] B. Walch, C. L. Cocke, R. Voelpel, and E. Salzborn, Phys. Rev. Lett. **72,** (1994) 1439.
[23] S. Martin, L. Chen, A. Denis, and J. Desesquelles, Phys. Rev. A **57,** (1998) 4518.
[24] J. P. Briand, L. d. Billy, J. Jin, H. Khemliche, M. H. Prior, Z. Xie, M. Nectoux, and D. H. Schneider, Phys. Rev. A **53,** (1996) R2925.
[25] A. Barany and C. J. Setterlind, Nucl. Instrum. Methods Phys. Res. B **98,** (1995) 184.
[26] T. Schlathölter, R. Hoekstra, and R. Morgenstern, J. Phys. B **31,** (1998) 1321.
[27] U. Thumm, J. Phys. B. **27,** (1994) 3515.
[28] U. Thumm, J. Phys. B. **28,** (1995) 91.
[29] U. Thumm, T. Bastug, and B. Fricke, Phys. Rev. A **52,** (1995) 2955.
[30] U. Thumm, A. Bárány, and H. Cederquist, Phys. Rev. A **56,** (1997) 4799.
[31] U. Thumm, Physical Review A **55 (1),** (1997) 479.
[32] J. Opitz, H. Lebius, B. Saint, S. Jacquet, B. A. Huber, and H. Cederquist, Phys. Rev. A **59,** (1999) 3562.
[33] N. Selberg, A. Barany, C. Biedermann, C. J. Setterlind, H. Cederquist, A. Langereis, M. O. Larson, A. Wännström, and P. Hvelplund, Phys. Rev. A **53,** (1996) 874.

Chapter 7

ELECTRON COLLISIONS

Electron impact ionisation of fullerenes is one of the most common ways of producing a beam of fullerenes for a variety of collision experiments but is also interesting in its own right. The ionisation/fragmentation mass spectra that are produced on electron impact ionisation show many similarities with those produced by highly charged ions or by interaction with ultrashort laser pulses. The characteristic feature of the mass spectra is the presence of multiply charged fullerenes [1-4]. The extraordinary stability of the fullerene structure is manifested in the relative ease with which non-fragmented multiply-charged molecules can be produced and detected. The detection of electron-impact produced C_{60}^{q+} with $q = 5$ [3], 6 [5] and 7 [6] was reported in rapid succession in 1994. Appearance energies for the different charge states have been reported [7] and are in good agreement with photoionisation values [8] and calculations [9], see also Chap. 3.2.1. The fragmentation behaviour of fullerenes following electron impact excitation is very similar to that of atom or ion impact excitation and can basically be described in terms of the statistical models and phenomenological fission models we have discussed in previous chapters. We will not repeat this here but instead will focus on the electron impact cross section data for both ionisation/fragmentation and electron attachment.

1. ELECTRON IMPACT IONISATION CROSS SECTIONS

Absolute partial and total cross sections were reported by a number of groups for electron impact ionisation of C_{60} [10-13]. The original cross sections reported by Dünser et al. [10] were later corrected to account for the occurrence of metastable fragmentation in the field free region of the mass spectrometer [14]. Some of the original and corrected data from the Märk group are shown in Fig. 7.1. Fig. 7.2 compares the measured (corrected) cross sections for single ionisation of neutral C_{60} measured by the different groups along with two theoretical calculations. Curve 1 is a calculation using the semi-empirical Deutch-Märk formalism [15]. The second theoretical curve (curve 6) is based on the same approach but uses a modified additivity concept [14]. The calculation uses a Mulliken population analysis of the fullerene orbitals in which the outermost 120 C_{60} orbitals are treated as linear combination of the atomic C(2s) and C(2p) orbitals. The calculated curves are significantly larger than the experimental cross sections and also are not able to reproduce the shape of the experimental curves More sophisticated theoretical models are obviously needed to satisfactorily explain the experimental measurements.

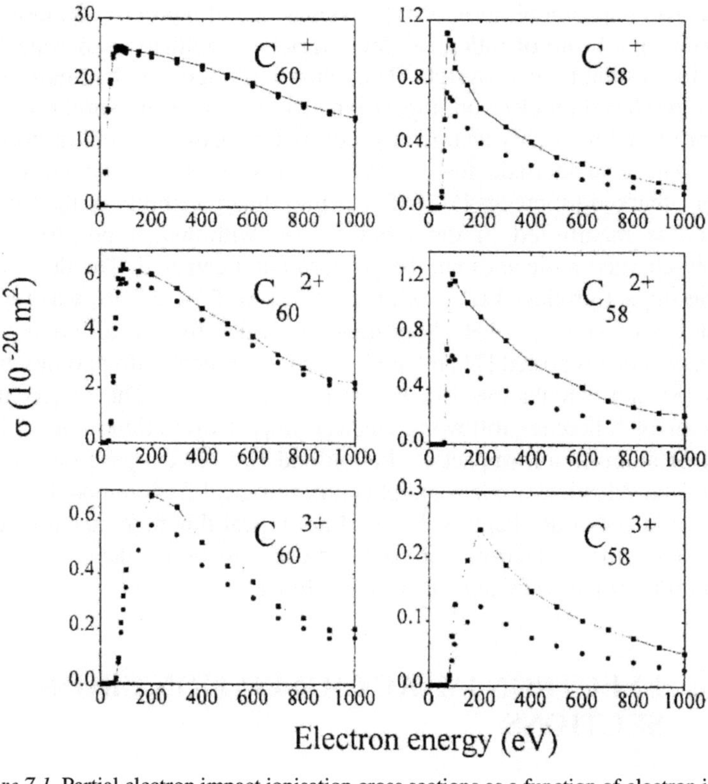

Figure 7.1. Partial electron impact ionisation cross sections as a function of electron impact energy. Circles from [11]. Squares from [14]. From [14].

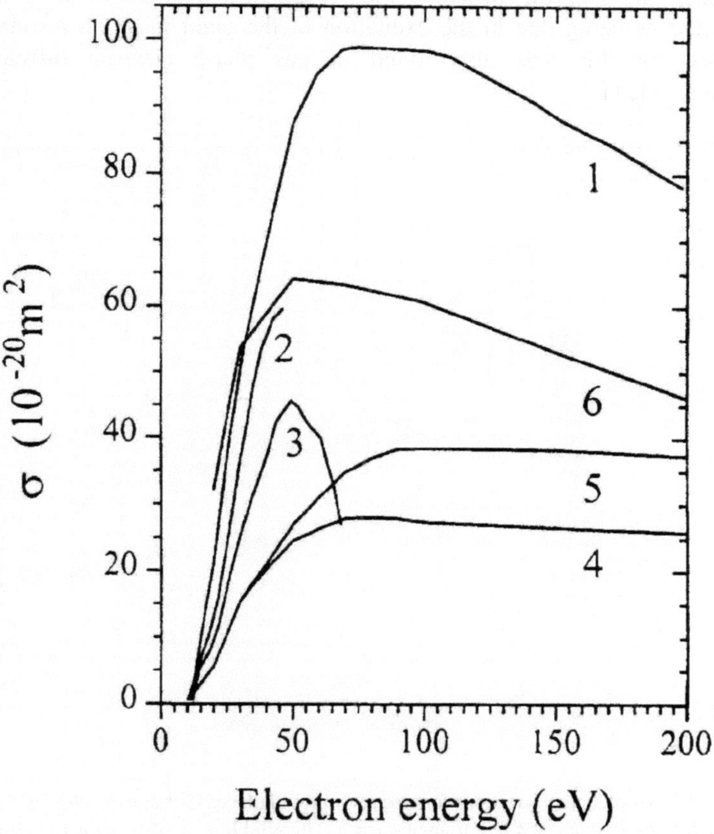

Figure 7.2. C_{60} total single ionisation cross sections. 1: calculated, from [15]. 2: experimental from [12]. 3: [13]. 4: experimental from [14]. 5: experimental from [11] corrected by [14]. 6: theoretical from [14]. From [14].

Cross sections have also been obtained for electron impact induced fragmentation and ionisation of fullerene ions [16-19]. the apparatus used for these studies is shown in Fig. 2.17. Absolute cross sections for electron impact induced C_2 fragmentation of $C_{60}{}^{q+}$ ions (q = 1,2,3) are shown in Fig. 7.3 [17]. The shape of the cross sections is very similar to those for collision induced dissociation of positively charged fullerenes in collisions with atomic targets (Fig. 5.5), however the time window in the experiment is quite different and the interpretation of the significant peak that appears with its maximum around 30 eV is also quite different. Here, the high energy part of the cross sections was fitted with the Lotz formula [20], normally used to calculate electron impact ionisation cross sections. This scales linearly with the charge state of the fullerene ions (Fig. 7.3). The dominant peak is interpreted as being due to the excitation of the giant plasmon resonance. Evidence for this was also found in gas phase electron diffraction experiments [21].

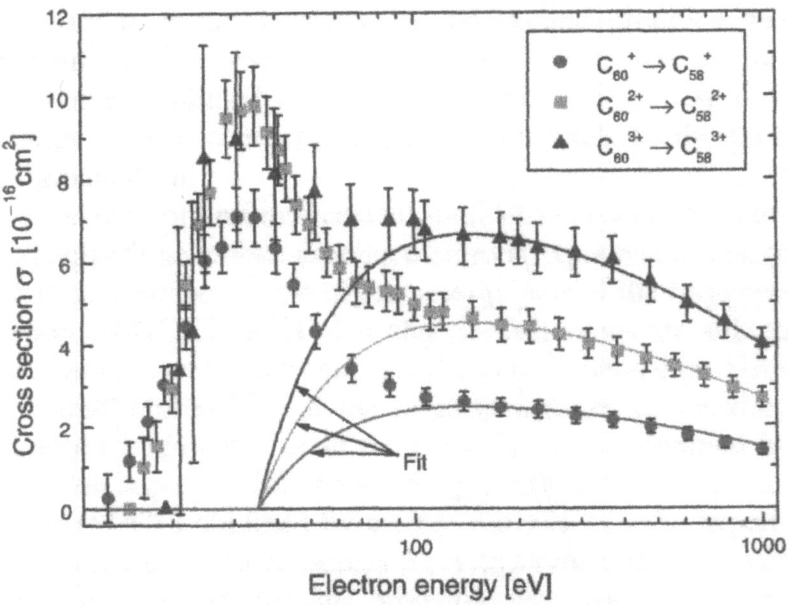

Figure 7.3. Absolute cross sections for electron impact induced C_2 fragmentation of C_{60}^{q+} ions. Circles: $q = 1$; squares: $q = 2$; triangles: $q = 3$. The solid lines are fits of the Lotz formula through the data points at energies higher than 120 eV. From [17].

A considerable effort has been made to study absolute cross sections for electron impact ionisation and fragmentation of negatively charged fullerene targets [16,18,19]. Cross sections for the double, triple and quadruple ionisation of C_{60}^-, C_{70}^- and C_{84}^- are shown in Fig. 7.4 [16]. There is very good agreement between the experimentally determined ratios of the cross sections for the double ionisation and the ratios of the geometrical cross sections. The triple ionisation data fit well with the squared ratios of the geometric cross sections and the ratios for quadruple ionisation match the cube of the ratios of the geometric cross sections. This can be described in the form of a scaling law [16]

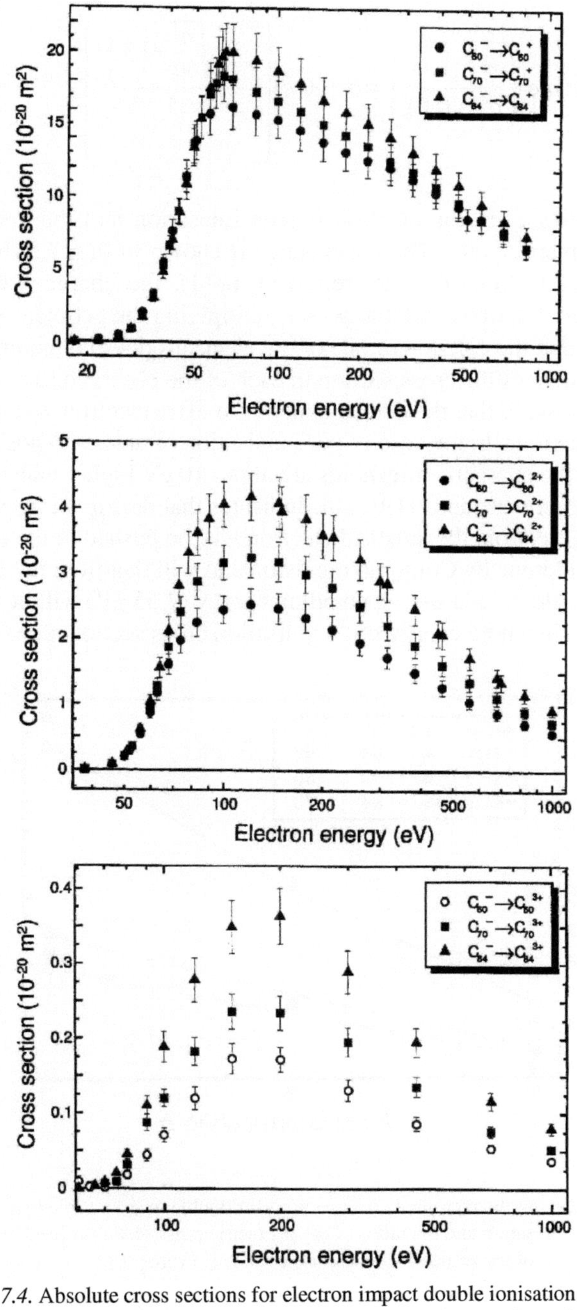

Figure 7.4. Absolute cross sections for electron impact double ionisation (upper), triple ionisation (middle) and quadruple ionisation (lower). From [16]

$$\sigma_e(n,q)=\sigma_e(60,q)\left(\frac{\sigma_{geom}(n)}{\sigma_{geom}(60)}\right)^q=\sigma_e(60,q)\left(\frac{\left(3.5\sqrt{\frac{n}{60}}+1.5\right)^2}{25}\right)^q \tag{7.1.1}$$

where σ_e and σ_{geom} are the electron impact ionisation and the geometrical cross section, respectively. The dependence is shown in Fig. 7.5. Instead of the number of electrons that are removed, $(q+1)$, the charge state of the product ion q corresponds with the power multiplying the geometrical ratios. This implies that the ejection of one of the electrons does not contribute to the magnitude of the cross section in each of the observed ionisation processes. It is likely that the detachment of the extra electron is responsible for this behaviour. In the case of pure ionisation of anions to positively charged ions, the energetic thresholds are about 10 eV higher than expected on purely energetic grounds [19]. This indicates that during the approach of the projectile electron, the attached electron will be pushed to the opposite side of the fullerene by Coulomb repulsion and will therefore not interact with the projectile. In addition, its binding energy (2.65 eV) will be reduced. Close to the threshold energy of an ionisation cross section all electrons

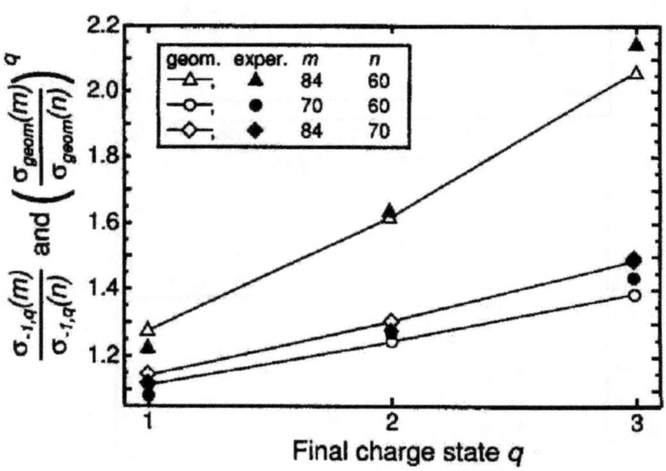

Figure 7.5. ratios of the measured cross sections for double (q = 1), triple (q = 2) and quadruple (q = 3) ionisation and the ratios of the geometric cross sections raised to the power of the respective final charge states q. From [16].

leaving the ion are slow. There will therefore be a relaxation of the formerly attached electron rather than its removal. However, at sufficiently large energies above the threshold, one of the emitted electrons which is normally ascribed to the former projectile electron will have enough energy to push the weakly bound electron completely away from the fullerene. On its way through the fullerene a fast projectile electron will most likely interact with the π-orbitals. As the number of electrons is so large, the removal of only a few electrons can be assumed to have only a little effect on the size of the orbitals. Therefore the cross scetion for the removal of q electrons from the π-orbitals can be expected to scale with the size of the π-orbital cloud to the power q and thus also with the size of the fullerene to the same power [16]. The scenario is illustrated schematically in Fig. 7.6.

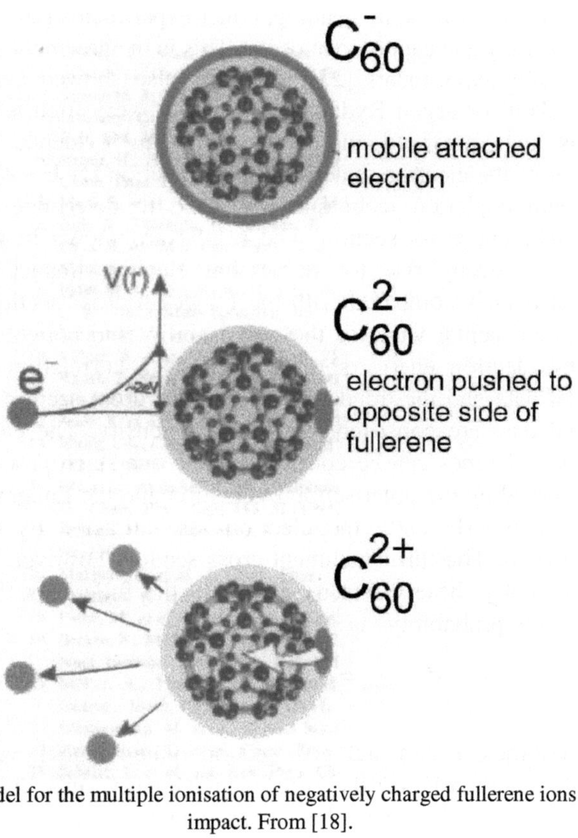

Figure 7.6. Model for the multiple ionisation of negatively charged fullerene ions by electron impact. From [18].

2. ELECTRON ATTACHMENT

Fullerenes have high electron attachment cross sections for low-energy electrons with a long plateau in the cross section for attachment to C_{60} existing until energies of around 10 eV. There has been a lot of discussion in the literature about the electron attachment cross sections. The results of early experiments [22-27] indicated that there was a barrier for attachment at zero incident energy and thus no s-wave capture. This was rationalised theoretically by Tosatti and Manini [28] (admittely with some rather crude approximations) and the apparent activation energy for attachment was attributed to a centrifugal barrier to electron capture in the p-wave channel of about 0.24 eV, in excellent agreement with experiment [23]. However, this has since been shown to be misleading. Other experiments [29,30] gave clear indications that s-wave capture can occur. This is in agreement with Rydberg electron transfer experiments [22,31,32] that also showed s-wave capture with either alkali or argon Rydberg atoms up to very high *n.* More recent experiments have provided additional convincing evidence for a strong enhancement of the electron attachment cross section for $E \rightarrow 0$ [33]. In this work, a beam depletion technique was used to determine the absolute integral attachment cross section. The results are shown in Fig. 7.7. The points show a strong rise for decreasing electron impact energy. The experimental data is compared with the Langevin cross section, calculated using the experimental value of the polarisability and convoluted with the experimental electron energy spread. There is a correlation between the experimental data and the trend in the Langevin cross section although the experimental data are consistently lower in intensity. The capture of free electrons by fullerenes can be considered as a two-stage process. The first stage is governed by the polarisation interaction and the Langevin curve is a measure of how efficiently the electrons are attracted by the particle's polarisation field. The full attachment cross section, however, also depends on how efficiently these electrons "stick" to the fullerenes [33]. One can define a sticking probability via

$$\sigma_{attahment} = \sigma_{Langevin} P. \tag{7.2.1}$$

P, the ratio of the experimental cross section to the Langevin value is clearly considerably less than 1 over the experimental energy range and shows some interesting structure. The results have consequences for modeling the reverse process of thermionic electron emission from fullerene anions.

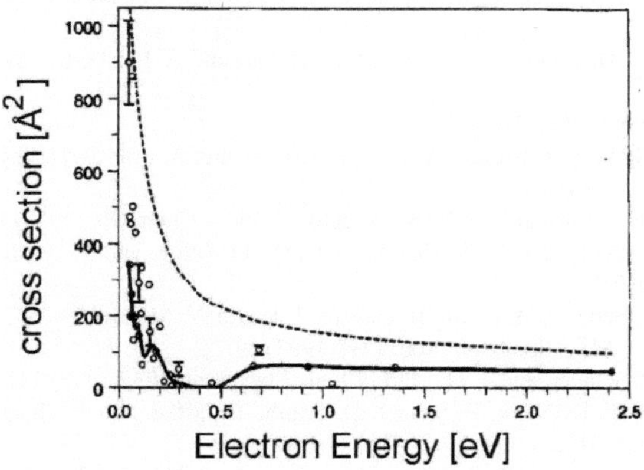

Figure 7.7. Experimentally determined integral cross sections for free electron attachment to C_{60}. Dashed line: Langevin polarisation capture cross section. Adapted from [33].

REFERENCES

[1] B. Dünser, P. Scheier, and T. D. Märk, Chem. Phys. Lett. **236,** (1995) 271.
[2] S. W. McElvany, M. M. Ross, and J. H. Callahan, Acc. Chem. Res. **25,** (1992) 162.
[3] P. Scheier, R. Robl, B. Schiestl, and T. D. Märk, Chem. Phys. Lett. **220,** (1994) 141.
[4] R. Völpel, G. Hofmann, M. Steidl, M. Stenke, M. Schlapp, R. Trassl, and E. Salzborn, Phys. Rev. Lett. **71,** (1993) 3439.
[5] P. Scheier and T. D. Märk, Int. J. Mass Spec. Ion Proc. **133,** (1994) L5.
[6] P. Scheier and T. D. Märk, Phys. Rev. Lett. **73,** (1994) 54.
[7] S. Matt, O. Echt, R. Wörgötter, V. Grill, P. Scheier, C. Lifshitz, and T. D. Märk, Chem. Phys. Lett. **264,** (1997) 149.
[8] H. Steger, J. Holzapfel, A. Hielscher, W. Kamke, and I. V. Hertel, Chem. Phys. Lett. **234,** (1995) 455.
[9] G. Seifert, R. Gutierrez, and R. Schmidt, Phys. Lett. A **211,** (1996) 357.
[10] B. Dünser, M. Lezius, P. Scheier, H. Deutsch, and T. D. Märk, Phys. Rev. Lett. **74,** (1995) 3364.
[11] S. Matt, B. Dünser, M. Lezius, H. Deutsch, K. Becker, A. Stamatovic, P. Scheier, and T. D. Märk, J. Chem. Phys. **105,** (1996) 1880.
[12] M. S. Baba, T. S. L. Narasimhan, K. Balasubramanian, and C. K. Mathews, Int. J. Mass Spectrom. Ion Proc. **114,** (1992) R1.
[13] A. A. Vostrikov, D. Dubov, Y,, and A. A. Agarkov, Techn. Phys. Lett. **21,** (1995) 715.
[14] V. Foltin, M. Foltin, S. Matt, P. Scheier, K. Beker, H. Deutsch, and T. D. Märk, Chem. Phys. Lett. **289,** (1998) 181.
[15] H. Deutsch, K. Becker, J. Pittner, V. Bonacic-Koutecky, S. Matt, and T. D. Märk, J. Phys. B **29,** (1996) 5175.
[16] D. Hathiramani, P. Scheier, K. Aichele, W. Arnold, K. Huber, and E. Salzborn, Chem. Phys. Lett. **319,** (2000) 13.
[17] D. Hathiramani, K. Aichele, W. Arnold, K. Huber, E. Salzborn, and P. Scheier, Phys. Rev. Lett. **85,** (2000) 3604.

[18] D. Hathiramani, P. Scheier, W. Arnold, K. Huber, and E. Salzborn, Physica Scripta T **92,** (2001) 153.
[19] P. Scheier, D. Hathiramani, W. Arnold, K. Huber, and E. Salzborn, Phys. Rev. Lett. **84,** (2000) 55.
[20] W. Lotz, Z. Phys. **206,** (1967) 205.
[21] L. G. Gerchikov, P. V. Efimov, V. M. Mikoushkin, and A. Solov'yov, Phys. Rev. Lett. **81,** (1998) 2707.
[22] J. Huang, J. H.S. Carman, and R. N. Compton, J. Phys. Chem. **99,** (1995) 1719.
[23] T. Jaffke, E. Illenberger, M. Lezius, S. Matejcik, D. Smith, and T. D. Märk, Chem. Phys. Lett. **226,** (1994) 213.
[24] M. Lezius, P. Scheier, M. Foltin, B. Dünser, T. Rauth, V. M. Akimov, W. Krätschmer, and T. D. Märk, Int. J. Mass Spec. Ion Proc. **129,** (1993) 49.
[25] M. Lezius, P. Scheier, and T. D. Märk, Chem. Phys. Lett. **203,** (1993) 232.
[26] S. Matejecik, T. D. Märk, P. Spanel, D. Smith, T. Jaffke, and E. Illenberger, J. Chem. Phys. **102,** (1994) 2516.
[27] D. Smith, P. Spanel, and T. D. Märk, Chem. Phys. Lett. **213,** (1993) 202.
[28] E. Tosatti and N. Manini, Chem. Phy. Letters **223,** (1994) 61.
[29] A. A. Vostrikov, D. Y. Dubov, and A. A. Agarkov, Tech. Phys. Lett. **21,** (1995) 517.
[30] O. Elhamidi, J. Pommier, and R. Abouaf, J. Phys. B **30,** (1997) 4633.
[31] C. D. Finch, R. A. Poppel, P. Nordlander, and F. B. Dunning, Chem. Phys. Lett. **244,** (1995) 345.
[32] J. M. Weber, M.-W. Ruf, and H. Hotop, Z. Phys. D **37,** (1996) 351.
[33] V. Kasperovich, G. Tikhonov, and V. V. Kresin, Chem. Phys. Lett. **337,** (2001) 55.

Chapter 8

ENDOHEDRAL FULLERENE FORMATION

Endohedral fullerenes (from the Greek "endo" meaning within) are fascinating molecules where one or more atoms are "captured" inside the carbon cage. An entire volume of the series "Developments in Fullerene Science" has been devoted to the production and characterisation of these novel materials [1]. The name "endohedral fullerene" was coined by Helmut Schwarz, the first person to show that an atom could be captured inside an intact carbon cage during a gas phase collision [2]. The idea of filling the fullerene cage is almost as old as fullerene research itself. Shortly after the discovery of buckminsterfullerene [3], evidence was found by the Rice group that lanthanum atoms can be trapped inside these empty carbon cages in experiments where a $LaCl_3$-doped graphite target was laser vapourised [4]. In the corresponding mass spectra, mass peaks according to LaC_n, $44 \leq n \leq 76$, were observed. The fact that in these compounds only even numbers of carbon atoms occurred was interpreted as a consequence of the presence of fullerene-like carbon cages surrounding the La atom. This hypothesis was later confirmed in "shrink-wrapping" experiments where the carbon cage was photofragmented down to the smallest size still capable of containing the metal atom [5]. Atom-containing fullerenes were given the appealing designation $M@C_n$. A number of different endohedrally metal-doped fullerenes could be synthesised by laser vapourisation [6, 7]. In addition to the shrink-wrapping technique, other experiments concerning the reactivity of the metal-fullerene compounds in the gas phase provided convincing evidence for the endohedral nature. Reactions of endohedral compounds with molecules such as O_2, NH_3, N_2O or H_2O do not occur in the gas phase whereas externally ("exohedrally") bound metal-fullerene species are readily oxidised [5, 8].

Other, more-efficient methods for the production of endohedral fullerenes were also developed. Using metal or metal oxide doped graphite rods it is possible to produce macroscopic amounts of endohedral fullerenes using the Krätschmer-Huffman arc discharge method for fullerene production [1, 9]. Although many different endohedral fullerenes can be observed in the mass spectra of the raw soot produced by this method, only the most stable compounds can be isolated and purified. In particular, it has proved very difficult to isolate $M@C_{60}$ using this method although some progress has been made in recent years [10, 11].

In this Chapter we will focus on the collisional formation of endohedral fullerenes, both in the gas phase and on substrates. For details of purification and characterisation of the more conventional and readily available arc-discharge produced materials, the reader is referred to the literature [1, 9].

1. GAS PHASE COLLISIONAL CAPTURE OF ATOMS AND IONS

1.1 Capture of Rare Gas Atoms

The collisional production of endohedral fullerenes in single gas-phase collisions was first reported by Weiske et al. [12] shortly after fullerenes became generally available in the early nineties. These experiments involved high-energy (8 keV laboratory collision energy, corresponding to a centre-of-mass energy of 44 eV) collisions between C_{60}^+ ions and neutral rare gas (RG) atoms. The apparatus used for an extensive series of studies is shown in Fig. 2.9. The energetic fullerene ions collided with helium atoms in the central field-free region of the four-sector mass spectrometer. Apart from the usual appearance of fragment ions from C_n^+, the collision-induced fragmentation pattern of C_{60}^+ also showed a second distribution of mass peaks shifted upwards by four mass units compared to the pure carbon fragments. The observation was interpreted as being evidence for the formation of an endohedral $He@C_{60}^+$ complex since an exohedrally bound HeC_{60}^+ ion should lose the loosely bound He atom before it undergoes loss of tightly bound C_2 molecules. Calculations indicated that the collision energy was sufficient to force a helium atom through a six-membered ring of the carbon cage [13, 14]. The experimental result was reproduced shortly afterwards by two different groups [15, 16] using three- and four-sector mass spectrometers. Figure 8.1 displays two mass spectra that were obtained by Ross and Callahan for collisions between C_{60}^+ and He at 44 eV (centre-of-mass) [16]. The upper spectrum displays the product distributions from these collisions. Apart from the primary ions, C_{60}^+, endohedral product ions,

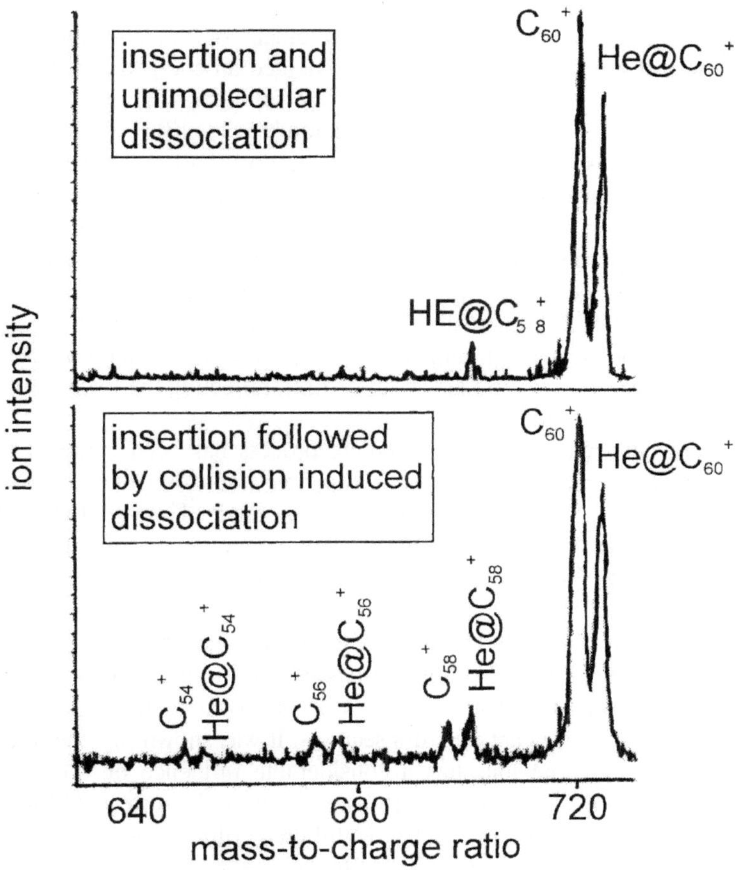

Figure 8.1. Three-sector mass spectra of collisions between C_{60}^+ and He at 44 eV (centre-of-mass). Adapted from [16]. Top: product distribution after unimolecular loss of C_2 in the field-free region of the mass spectrometer. Bottom: product distribution after collisions with Xe atoms at 200 eV. The fact that the He complexes survive the second collisions provides evidence for their endohedral character.

$He@C_{60}^+$ (mass 724u) and fragment ions C_{58}^+ and $He@C_{58}^+$ are visible. Evidence for the endohedral character of the products can be seen in the lower part of Fig. 8.1 that shows the result of an experiment where the primary collision products were collided with Xe atoms at 200 eV (centre-of-mass). The He-carbon complexes evidently survive these collisions, which can only be explained if capture of He inside the fullerene cage is assumed.

Another clear indication of the endohedral nature was obtained in time-of-flight mass spectrometry experiments. In these experiments it was much more straightforward to change the collision energy than in the complex commercial multi-sector machines. Measurement of the collision energy dependence of the formation of $He@C_{60}^+$ yielded an energetic threshold of approximately 6 eV for He to penetrate the fullerene cage [17], in good agreement with predictions for ring penetration [13, 14, 18].

In subsequent work by the Schwartz group, it was shown that He can also be inserted into doubly and triply charged fullerene ions [19] and that sequential insertion of ^3He and ^4He can occur in separate collisions [20]. One of the nicest experiments of this time was the illustration of the stability of the endohedral complex by carrying out neutralisation-reionisation collisions, Fig. 8.2 [21]. Interestingly, no direct capture products $He@C_{58/68}^+$ but only their post-capture fragments $He@C_n^+$, $n < 58,68$ were observed in collisions between $C_{58/68}^+$ and He at 44 eV [22], reflecting the lower stability of $C_{58/68}^+$ ions relative to $C_{60/70}^+$.

A deeper understanding of the capture process has been obtained from measurements of the capture cross sections as a function of the collision energy. Two different methods have been used. Time-of-flight mass spectrometry was used by the author's group in order to investigate the capture of He and Ne atoms by positively and negatively charged fullerene ions [17, 23-25]. The Anderson group used a triple-sector guided ion beam mass spectrometer to study capture of Ne^+ ions in neutral C_{60} [26, 27]. In contrast to the majority of studies discussed so far in this chapter, these experiments allowed the clear determination of the energetic thresholds for capture and of absolute capture cross sections. It was shown that the internal energy of the fullerene ions has a considerable influence on the capture threshold. Different ion sources were used to study this effect. Using C_{60}^+ ions produced by laser desorption from a fullerene film yielded thresholds of 6±2 eV and 9±1 eV for the capture of He and Ne, respectively [17]. These values increase by approximately 1.5 eV if negatively charged C_{60}^- projectile ions, produced by electron attachment in an electron impact ionisation source are used [24], Fig. 8.3. Negative fullerene ions produced in this way have lower internal energies than do hot positive ions from a laser desorption or electron impact ionisation source because electron attachment occurs efficiently only for electron energies below 10 eV [28]. This internal energy effect on the capture threshold has also been confirmed by classical MD simulations [29]. The simulations implied that the internal energy of the laser desorbed positively charged ions should be approximately 30 eV higher than that for the negative ions produced by electron attachment in order to explain the shift in capture threshold observed in the experiments. The excitation energy of C_{60}^- can be estimated to be 7.5 eV (thermal energy from the oven plus the electron affinity) which would then imply an internal

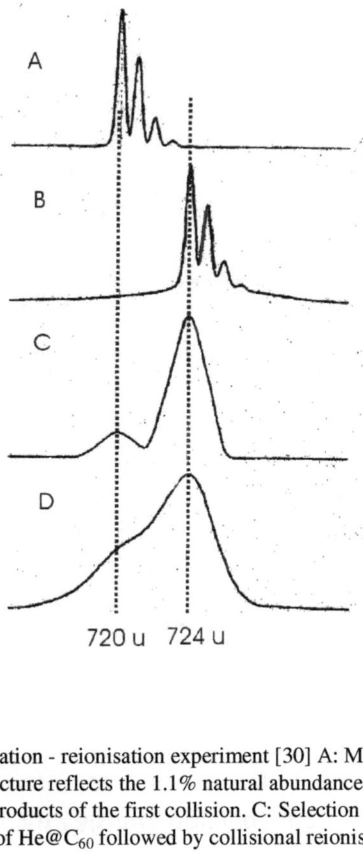

Figure 8.2. Neurtralisation - reionisation experiment [30] A: Mass -resolved C_{60}^+ ions. The multiple peak structure reflects the 1.1% natural abundance of ^{13}C B: mass-selected endohedral He@C_{60} products of the first collision. C: Selection of $^4He@^{12}C_{60}$. D: Ion signal after neutralisation of He@C_{60} followed by collisional reionisation. Adapted from [30].

energy of approximately 35 eV for C_{60}^+. This can be compared with the estimate of 30-40 eV based on the amount of metastable fragmentation observed from the C_{60}^+ parent ions from the source in these experiments and the fragmentation rate constant, discussed in Chap. 3.

The decreasing capture threshold with increasing internal energy reflects the higher flexibility of an excited fullerene molecule. The effective size of the hexagons in the fullerene cage increases and therefore leads to a decrease of the penetration barrier [29]. The measured He capture threshold decreases even down to as low as 3 eV if C_{60}^- projectiles from a laser desorption source with high fluence (and thus high internal energy) are used in the collision

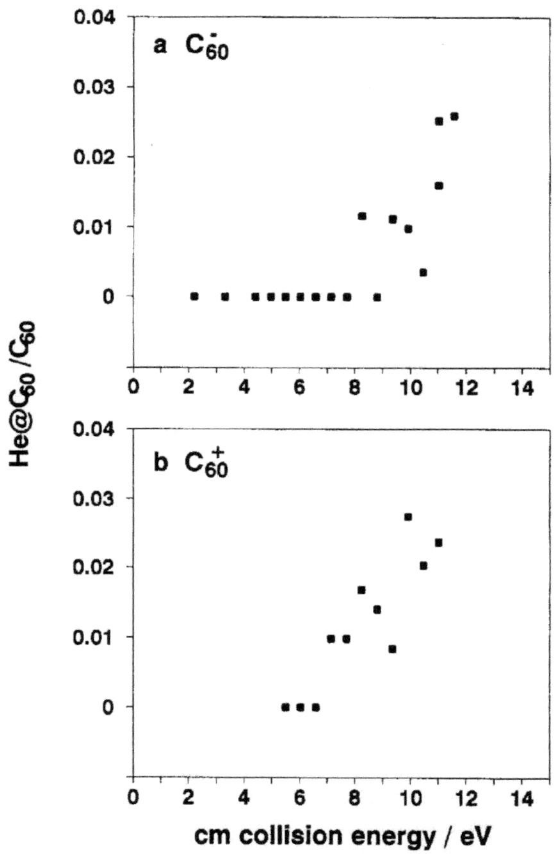

Figure 8.3. Cross sections for the production of endohedral He@C_{60}. a) negatively charged C_{60}^- (b) positively charged C_{60}^+. The different threshold can be attributed to the initial internal energy of the fullerene ions. From [24].

experiment [25]. This value is extremely low, however, a 3 eV escape threshold has also been estimated from high pressure experiments (discussed in section 8.1.2, below). A possible explanation for these observations was provided by Murry and Scuseria [31]. In their semi-empirical and ab initio calculations they observed the formation of thermodynamically stable windows on the potential energy surface of the triplet manifold of C_{60} corresponding to nine- and ten-membered rings in the fullerene cage, Fig.

8.4. Such "windows" have also been seen in MD simulations of C_{60} with an internal energy of approximately 40 eV, due to the high propensity for defect formation in the cage at such high excitation energies [32]. This, combined with calculations that show that annealing processes are much less efficient with anions than with cations [33], could explain the reduced capture threshold for fullerene anions with very high initial internal excitation energies in the collision experiments [25]. It also provides a plausible explanation for the rather low value determined for capture of Ne using laser-desorbed fullerenes [17].

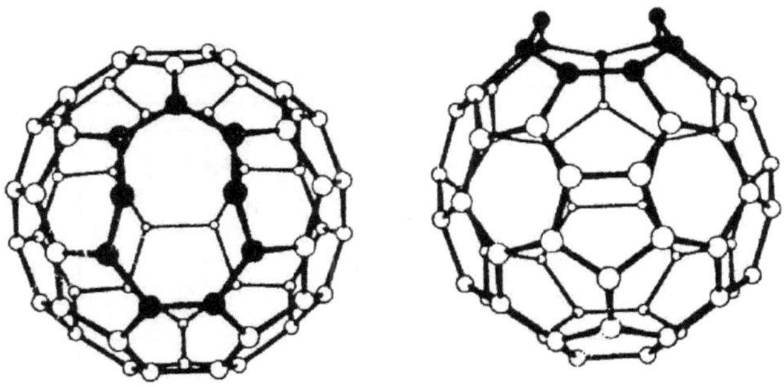

Figure 8.4. Window on the potential energy surface of the triplet manifold of C_{60} suggested as an explanation for the high pressure encapsulation of rare gas atoms in C_{60}. Adapted from [31].

The collision energy dependence of the reaction products in collisions between Ne^+ and C_{60} was reported by Christian et al. [26]. Fig. 8.5 shows the relative cross sections for collision-induced fragmentation (top) and capture followed by fragmentation (bottom) as a function of collision energy. The formation of these ions is accompanied by charge transfer from the neutral C_{60} target to the Ne^+ ions (note that this process is exoergic by 13.95 eV). The charge transfer cross section, represented by the C_{60}^+ ion signal in the upper part of the figure, is approximately a factor of 100 times larger than the cross section for the reactive collisions. The shapes of the C_n^+ and $Ne@C_n^+$ ion signals are very similar. With increasing collision energy, the intensity goes from a threshold region through a maximum after which it decreases. The positions of threshold and maxima shift to higher energies with decreasing fragment size, n. The threshold shift, however, is not present

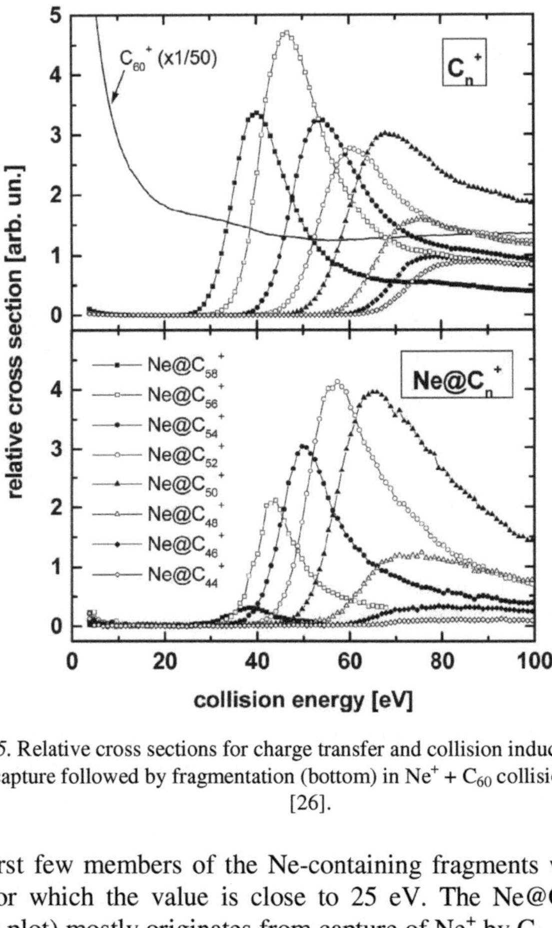

Figure 8.5. Relative cross sections for charge transfer and collision induced fragmentation (top) and capture followed by fragmentation (bottom) in Ne^+ + C_{60} collisions. Adapted from [26].

for the first few members of the Ne-containing fragments with n = 60, 58 and 56 for which the value is close to 25 eV. The $Ne@C_{60}^+$ signal (not shown on plot) mostly originates from capture of Ne^+ by C_{70} and subsequent fragmentation (15% of the target gas consisted of C_{70}) since most of the primary $Ne@C_{70}^+$ fragment on their way to the detector. Note that the flight times in these experiments are roughly three orders of magnitude longer than in the time-of-flight or three- or four-sector mass spectrometer experiments discussed above. The threshold energy of 25 eV for the detection of $Ne@C_{60}^+$ is in rough agreement (although somewhat high) with capture thresholds reported from other experiments and theory (see below), however, the electron transfer which occurs before the actual atom capture in this study may limit direct comparability. The similarity of the fragment patterns of pure carbon and Ne-containing ions demonstrates that evaporative cooling, essentially via loss of C_2 molecules, is basically the same for fullerenes and rare gas endohedrally doped fullerenes. This is also in agreement with measurements of kinetic energy releases in metastable fragmentation of endohedral $Ne@C_{60}^+$ [34].

There are actually two mechanisms for capturing He in collisions with fullerene ions. Figure 8.6 depicts relative cross sections for capture of He and Ne atoms by C_{60}^+ as a function of inverse collision energy [23]. An expression describing the reaction cross section in terms of a simple absorbing sphere or line-of-centres model [35] is fitted to the experimental data

$$\sigma_{capture} = \pi R_{12}\left(1-\frac{E_{thres}}{E}\right) \quad (8.1.1)$$

where E_{thres} is the energetic threshold for capture. The total energy during the collision can be expressed as the sum of kinetic energy T and potential energy $V(R)$, $E = T + V(R)$. The kinetic energy is the sum of radial and centrifugal contributions

$$T = \frac{1}{2}\mu\left(\frac{\partial R}{\partial t}\right)^2 + \frac{l^2h^2}{4\pi\mu R^2} = \frac{1}{2}\mu\left(\frac{\partial R}{\partial t}\right)^2 + \frac{b^2E}{R^2}, \quad (8.1.2)$$

where R is the intermolecular distance, l the rotational quantum number, μ the reduced mass and b the impact parameter. It follows that

$$E = \frac{1}{2}\mu\left(\frac{\partial R}{\partial t}\right)^2 + V(R) + \frac{b^2E}{R^2}. \quad (8.1.3)$$

The capture cross section can be obtained from the assumption that the intermolecular potential $V(R)$ with the barrier E_{thres} at the contact distance R_{12} of the collison partners, together with the centrifugal energy, has to be overcome by the radial kinetic energy

$$E - E_{thres} - \frac{Eb^2}{R_{12}^2} > 0. \quad (8.1.4)$$

The capture cross section (eq. 8.1) follows from the maximum impact parameter that satisfies this inequality. The same simple model is applied to fullerene-fullerene collisions in Chap. 9.

Two separate linear regimes are seen for He capture in Fig. 8.6, corresponding to two different capture mechanisms. The low energy capture for collisions with He is characterised by a penetration barrier of E_{thres}= 6±0.5 eV, in excellent agreement with earlier studies [17]. The slowly rising cross section as the collision energy increases indicates a very low capture

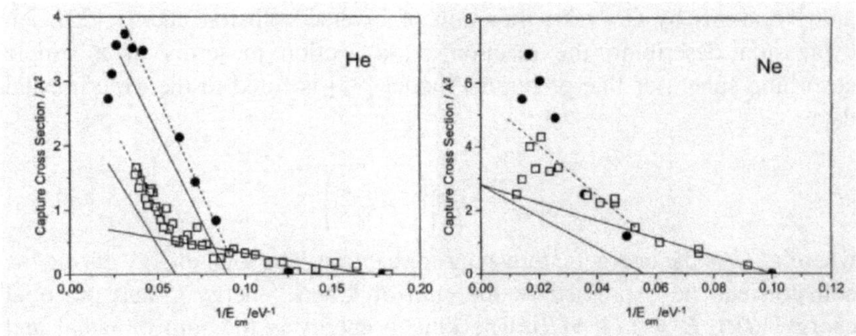

Figure 8.6. Capture cross sections for collisions between C_{60}^+ and He (left) or Ne (right) atoms [23]. Squares: experimental capture cross sections. Circles: results of MD simulations. Lines are fits of the line-of-centres model (eq. 8.1.1) to the data.

probability with R_{12} ≈0.5 Å. This indicates that only central collisions through a six-membered ring contribute to He capture at these energies. The threshold energy determined for the process that onsets at higher collision energies is E_{thres} = 17±0.5 eV. The steeper slope indicates a larger probability for capture via this second process.

In order to understand the mechanism for the high energy capture, we consider additional information provided from measurements of inelastic scattering in C_{60}^+ + He collisions. The upper part of Fig. 8.7 shows the experimental kinetic energy distribution of product ions from C_{60}^+ + He collisions at a centre-of-mass collision energy of 27.6 eV [18]. Similar data were reported in [36] and [23]. The sharp feature in the broad scan corresponds to the endohedral collision product as is evident from the narrow scan (below) which was taken with an additional mass filter set to 724 u before ion detection. The lower part of Fig. 8.7 shows the result of a corresponding classical MD simulation that agrees very well with the experimental result [23]. The low energy foot of the kinetic energy distribution does not correspond to fully elastic collisions with a kinetic energy loss in the laboratory system of approximately $4E_{cm}$ as would be expected for a very light target like He (Chap. 3.1). The energy loss in the

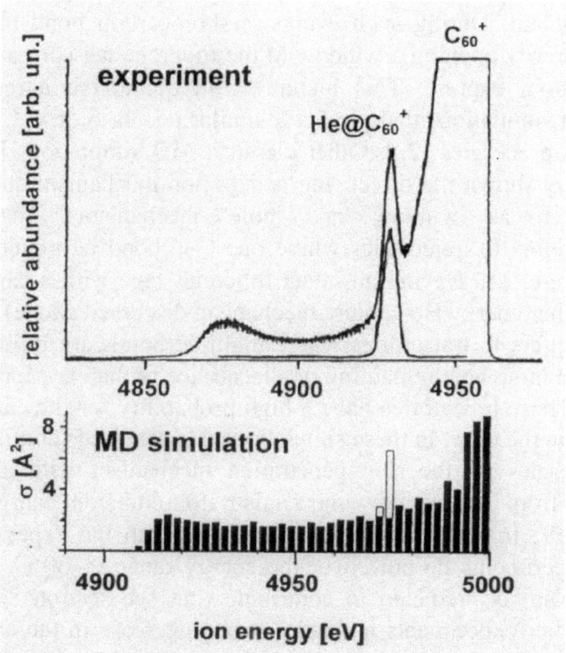

Figure 8.7. Top: kinetic energy scans of products from C_{60}^+ + He collisions at 4.95 keV (lab). Adapted from [18]. In the broad scan both C_{60}^+ and $He@C_{60}^+$ were transmitted to the detector while the narrow scan was obtained with an additional mass filter set to 724 u, corresponding to the endohedral helium complex. Bottom: kinetic energy distributions of C_{60}^+ projectiles after collisions with He at 5 keV (lab) calculated with classical MD simulations [23]. The white bar indicates the position of the kinetic energy of the endohedral complex $He@C_{60}^+$.

laboratory frame of reference is smaller and can be understood in terms of the spectator model discussed in Chap. 3.1.1. The He atom first collides fully elastically with only a part of the fullerene cage e.g. a C atom or a dimer, leading to the observed laboratory kinetic energy loss. Subsequently, this sub-unit on the fullerene cage that has been scattered in the collision, collides with the remaining fullerene cage, finally leading to a statistical distribution of vibrational excitation in the fullerene. The effective mass m_{eff} that the atom "sees" on the cage depends on the collision energy and can be calculated from the maximum of the kinetic energy loss spectrum [23]. This has been plotted in Fig. 3.4 and it can be seen that m_{eff} decreases rapidly for collision energies between 10 and 17 eV from 720 u to 24 u and remains constant close to the mass of a C_2 unit in the energy range where the most effective He capture mechanism is active. This indicates that for collision energies above 17 eV, direct knock-on collisions with C_2 sub-units on the cage are efficient. During such events, carbon-carbon bond breaking can occur, temporarily opening a window in the fullerene network and therefore facilitating atom capture. This picture is in qualitative agreement with classical MD simulations that predict a similar decrease of m_{eff}, however at lower collision energies [23]. Other classical MD simulations [15, 29, 37] have explicitly shown the direct ring penetration mechanism and also what are referred to as "window" and "hole" mechanisms. The "window" mechanism refers to trajectories where one C-C bond is broken on impact and rapidly re-closes leaving an intact fullerene cage with a captured atom (as in the higher energy He-capture mechanism discussed above). The "hole" mechanism refers to trajectories where multiple bonds are broken on atom impact with a large hole remaining on the surface of the cage for up to a few hundred fs. These trajectories have a high probability for the captured atom to escape from the cage. In these simulations, He capture is dominated at low collision energies by the ring penetration mechanism with a significant contribution from the window mechanism for collision energies beyond roughly 20 eV, in good qualitative agreement with the experiments (Fig. 8.6). The experiments do not cover the energy range (> 30 eV) where the hole mechanism is predicted to contribute (but see section 8.2.1 below), however, earlier experiments noticed a strong decrease in the capture cross section for collision energies beyond this value [17, 18]. This is also observed for collisions between C_{60} and Li^+ ions (see section 8.1.2, below).

The situation for Ne capture is rather different. Experimentally, there are also two mechanisms observable with highly excited laser desorbed fullerene ions (Fig. 8.6). The low-energy threshold at about 10 eV is attributed to penetration of the atom through defects, already present on the cage surface prior to collision. The higher energy process is due to a bond-breaking mechanism [23]. In the Anderson experiments, Fig. 8.5, the fullerenes are not so highly excited prior to collision and the low-energy capture mechanism is not observed. Taking into account the experimental errors and the differences in temperature in the two kinds of collision experiments, the threshold found by the Anderson group [27] is consistent with the high-energy threshold determined by the author's group [23]. Interesting behaviour is observed when one considers the scattering dynamics by analysing the effective mass seen by Ne, as discussed above for He. In contrast to He, the effective mass seen in the Ne experiments does not drop to 24 u but reaches a minimum of only about 60-100 u for collision energies beyond 30 eV [23]. This is consistent with the larger Ne atom colliding with a C_6-ring on the cage, inducing multiple bond-breaking and thus predominantly being captured by the hole mechanism as discussed by Cui et al. [37] and inferred by them to be the main capture mechanism for Ne.

The capture cross section drops with increasing collision energy (Fig. 8.5). This is due, partly, to fragmentation of the endohedral complex on the μs or ms time scale of the experiments. For small captives such as He or Li (discussed below) the loss of the guest atom may compete with the evaporation of C_2 molecules. This is particularly likely at high collision energies where there is a greater probability for the production of a metastable hole in the fullerene cage on impact. Secondly, at high collision energies, direct "shoot-through" trajectories are possible during which an atom enters the fullerene but is not trapped and leaves the cage some fs after its capture. Such trajectories have been observed in MD simulations [38, 39].

Ar-doped endohedral fullerenes have also been reported in collision experiments with the subsequent formation of $Ar@C_{60}^+$ fragments with an odd number of carbon atoms [15]. This observation rules out the standard C_2 loss fragmentation mechanism and indicates a direct collisional induced fragmentation rather than a statistical mechanism, although to the author's knowledge it has not been confirmed in other studies. Collision-induced fragmentation of pre-prepared $Ar@C_{60}^{+/-}$ leads to the typical even-numbered fragment distribution [40]. The larger Kr and Xe noble gas atoms cannot form metastable endohedral complexes in collisions under single-collision conditions. However insertion of these atoms into C_{60} is possible with the high-pressure technique of Saunders [41] or by using the nuclear recoil method [39, 42]. Both these methods are discussed below.

1.2 High Pressure Encapsulation of Rare Gas Atoms

The discovery that He atoms could be incorporated into fullerenes during the arc discharge production process was made by M. Saunders [43], however the yield was extremely low and the discovery was only made possible by using highly specialised mass spectrometry methods. Subsequently, the Yale group developed a high pressure - high temperature method of incorporating rare gas atoms into the fullerene cage [44]. In this method the fullerenes are heated to a temperature of 600 °C for a number of hours under a high pressure (2500 atm) of the rare gas atom. Analysis of the fullerene material after this treatment shows that approximately 0.03-0.1% of the fullerene molecules contain the rare gas atom. The presence of He inside the fullerene cage does not significantly affect its chemical or other properties. NMR experiments with endohedral fullerenes containing ^3He have therefore been used to monitor the magnetic field at the centre of the cage during chemical reactions [45]. The presence of the larger rare gas atoms, Ar and Kr, inside the cage does affect the cage properties slightly. Both $Ar@C_{60}$ [46] and $Kr@C_{60}$ [47] have been partially purified using HPLC (high pressure liquid chromatography). Although the UV-vis absorption spectrum of $Kr@C_{60}$ is very similar to that of C_{60}, it does show some significant differences such as a red shift of peaks in the ${}^1T_{1g} \rightarrow {}^1A_{1g}$ dipole forbidden band and a blue shift of the electronically allowed band at 400 nm.

1.3 Gas Phase Collisional Capture of Alkali Ions

The capture of alkali ions in neutral fullerenes is similar in many respects to the capture of rare gas atoms in charged fullerenes, discussed in section 8.1.1 above. The radii of the two smallest alkali ions (Li^+ and Na^+) are comparable to those of He and Ne and we may expect similarities in the dynamical behaviour on capture. Capture of alkali metal ions by fullerenes has been investigated in detail by the Anderson group [48, 49]. In their experiments, metal ion projectiles collide with neutral fullerene target molecules. Mass-separated collision products are detected as a function of the collision energy. The experiments are similar to the Ne^+ capture studies from the same group (discussed above in section 8.1.1) with the important difference that in the case of alkali metal ions no charge transfer is involved in the collision process due to the lower ionisation potentials of the alkali atoms compared to the fullerenes. At the low collision energies of these experiments endoergic charge transfer is of no significance. Figure 8.8 shows the relative cross sections of the different reaction channels in collisions between both Li^+ and Na^+ with C_{60} (the target gas cell contained 15% C_{70}). The intensity of the metal-doped parent ions $Li@C_{60}{}^+$ and $Na@C_{60}{}^+$ is plotted in Fig. 8.9 as a function of $1/E_{CM}$, as was done for He and Ne collisions, above. The parent ion signals behave in a very similar way to the rare gas capture collision products. Both sets of data show evidence of two collision mechanisms. The Li product has a low energy threshold at approximately 5.5 eV with a more efficient process onsetting at approximately 15 eV. As for He collisions, this can be related to the threshold for low impact parameter penetration of a six-membered ring and for capture via a C_2 bond-breaking mechanism. The C_6 penetration barrier for Li^+ is in good agreement with calculations [50, 51]. There does not appear to be a temperature dependence in the capture probability for Li^+, as shown in a detailed series of quasiclassical trajectory calculations [50]. This is in contrast to the situation with He [25, 29] and may reflect the smaller radius of Li^+ (0.2 Å compared with 0.3 Å for He). No experimental results on the temperature dependence are available for the alkali ion projectiles.

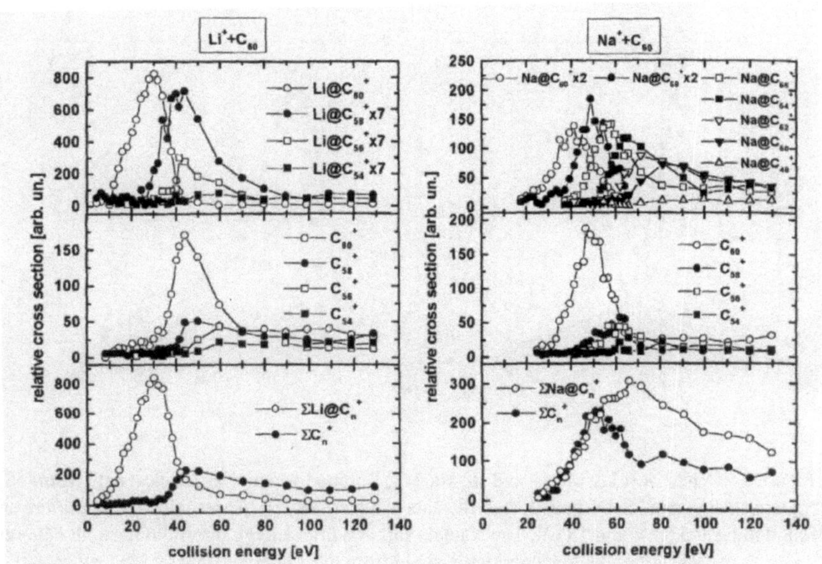

Figure 8.8. Relative cross sections for the formation of endohedral metallofullerenes in alkali ion - C_{60} collisions as a function of collision energy. Adapted from [48]. Left: Li^+; right: Na^+ projectile ions.

The $Na@C_{60}^+$ product has a similar collision energy dependence as the $Ne@C_{60}^+$. The first threshold lies at approximately 15 eV (corresponding to the C_2 bond-breaking also seen for Li^+). Na^+ is too large to penetrate an intact six-membered ring in the carbon cage. A second capture mechanism is seen to onset for a collision energy of approximately 28 eV. This can be related to the "hole" mechanism discussed in section 8.1.1. A more recent set of experimental results for $Na^+ + C_{60}$ [52] is also shown on Fig. 8.9(b) for comparison. The newer data was obtained using time-of-flight mass spectrometry in contrast to the ion guide method used for the earlier experiments. The newer measurements have a much shorter and better defined time window for detection of the products and also a better collision energy resolution [52]. The original reports gave threshold energies of ca. 18 eV [49] and 29.6 eV for Na^+ capture [52]. The discrepancy was attributed to the collision energy uncertainty in the earlier experiments. The plot of the two sets of data versus $1/E_{CM}$ show that in fact the two sets of data do not give conflicting threshold energies. Both data sets show a threshold at approximately 30 eV. The earlier data also show a lower energy capture threshold. The linear extrapolation places this threshold at approximately 15 eV, in agreement with the Li^+ measurements and the C_2 bond-breaking

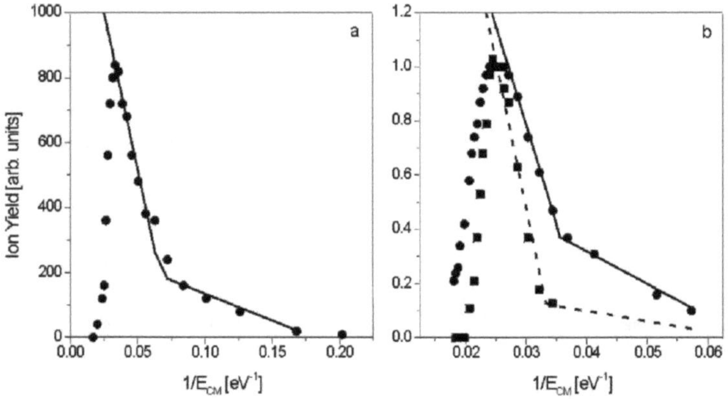

Figure 8.9. Yield of (a) $Li@C_{60}^+$ and (b) $Na@C_{60}^+$ plotted versus 1/E. Circles: data from [48]. Squares: data from [52]. Lines: Fits to the data using eq. (8.1.1). Thresholds for Li capture are found to be at 5.5 eV and 15 eV. The Na data shows a high energy threshold for both data sets of 29±1 eV and a low energy threshold from the Anderson data of 15±1 eV.

picture. It is not clear why the low energy mechanism is not observed in the later experiments perhaps it is simply a matter of sensitivity. The two data sets have been normalised to the same intensity maximum in Fig. 8.9. One would expect a larger fragmentation of the product ion in the guided ion beam experiment since there is a much longer time before detection (0.5-5 ms compared with 10 μs for the time-of-flight experiments). Thus the maximum in the parent endohedral ion intensity should be decreased relative to the low energy tail compared to the time-of-flight data. (For an internal energy of 40 eV, a rough estimate, taking account of radiative cooling, yields that ca. 90% of the parent endohedral ion should survive to be detected in the time-of-flight setup compared to ca. 70-80% in the guided ion beam experiment.) Of course, one cannot rule out the effect of the poorer collision energy resolution in the early experiments, and the different widths of the data in Fig. 8.9 show that this is indeed significant, but it is unlikely to be able to explain the full extent of the low energy tail in the guided ion beam data.

The initial internal energy of the collision partners and thus the total excitation energy after formation of the endohedral compound is much better defined in the alkali ion implantation experiments compared to experiments using fullerene ion beams. The fullerene target has an internal energy defined by the oven temperature with a corresponding thermal energy spread. The ion source has an energy spread of approximately 0.3 eV and the amount of energy transferred to internal degrees of freedom is just the centre-of-mass collision energy. Deng and Echt made use of this to estimate the binding energy of C_2 in $Na@C_{60}^+$ by modelling the collision energy dependence of the parent ion and the $Na@C_{58}^+$ fragment ion intensities [52]. By adopting a pre-exponential factor of A = 5 x 10^{19} s^{-1} (eq. 3.2.1) they obtained a value of D = 10.17±0.02 eV (only statistical error). This is consistent with the value of 10.3 eV for C_2 loss from C_{60}^+ derived by considering the competition between thermionic electron emission and fragmentation in the neutral channel [53], however this work fitted a pre-exponential factor of $3.4x10^{21}$ s^{-1}. If the pre-exponential factor was increased for $Na@C_{60}^+$ this would require a corresponding increase of D in order to still fit the experimental results (an increase of A by one order of magnitude would increase the derived binding energy by 0.62 eV [52]). A surprisingly large enhancement of the C_2 binding energy by ca. 23% for rare gas endohedrals $Rg@C_{60}^+$ (Rg = Ne, Ar, Kr), formed by the high pressure method discussed above, compared to C_{60}^+ was reported by Laskin et al. [54]. However, there are many difficulties associated with determining the binding energy from kinetic energy release measurements, as was done in this work, (see Chap 3.2.1) and one would need an independent confirmation to be sure that this is a genuine effect.

2. ION IMPLANTATION

2.1 Low Energy Ion Implantation

The gas phase experiments, discussed above, are related to a low energy ion implantation method developed as an alternative way to obtain macroscopic amounts of endohedral C_{60} compounds [55]. The experimental set up used to produce macroscopic amounts of alkali endohedral metallofullerenes is shown in Fig. 8.10. Alkali ions are produced from a thermal source, accelerated to the desired energy and deposited on a rotating metal cylinder with an ion beam cross-sectional area of approximately 2 cm^2 and a current of typically a few μA. The ion beam energy is controlled to within ± 1 eV. Fullerenes are deposited simultaneously from an oven at a temperature of approximately 450 °C. The oven temperature, fullerene deposition rate and speed of rotation of the metal cylinder are controlled to ensure that one monolayer of fullerenes is deposited on each full rotation.

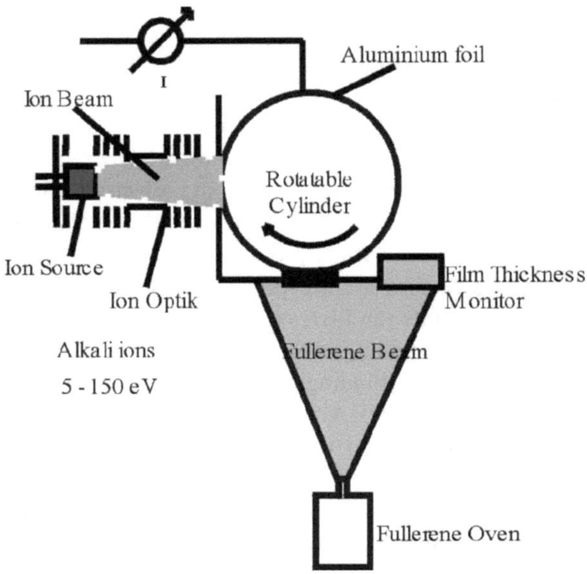

Figure 8.10. Schematic diagram of apparatus used to produce macroscopic amounts of $Li@C_{60}$ via low energy ion implantation.

The deposition and irradiation is continued until films of typically up to a few hundred-nm thickness are produced. For ease of extraction of the endohedral fullerene material, the fullerenes are deposited on aluminium foil wrapped around the cylinder. After a production run the aluminium foil is removed from the vacuum apparatus, cut into small pieces and put into CS_2 to dissolve the fullerenes.

The films produced by irradiation of fullerenes with alkali ions are routinely characterised by laser desorption time-of-flight mass spectrometry (LDMS) [56]. An example of an LDMS mass spectrum from a film prepared at 30 eV Li ion energy and a high ion current corresponding to approximately 17 Li^+ deposited per fullerene molecule is shown in Fig. 8.11. Under such conditions it is possible to produce also Li_2 and Li_3 endohedral species. However, films produced with such high ion currents are rather difficult to dissolve and for the routine material production it is more efficient to use a lower current corresponding to Li^+:C_{60} of 1:1. The LDMS method gives a lower limit to the efficiency of production of endohedral fullerenes. Since the molecules are desorbed with very high internal energies, corresponding to temperatures of a few thousand Kelvin, both thermal ionisation and fragmentation takes place on desorption [57]. The magnitude of the endohedral mass peak compared to empty fullerenes and fragments thus generally depends on the time scale of the ion extraction and detection [58].

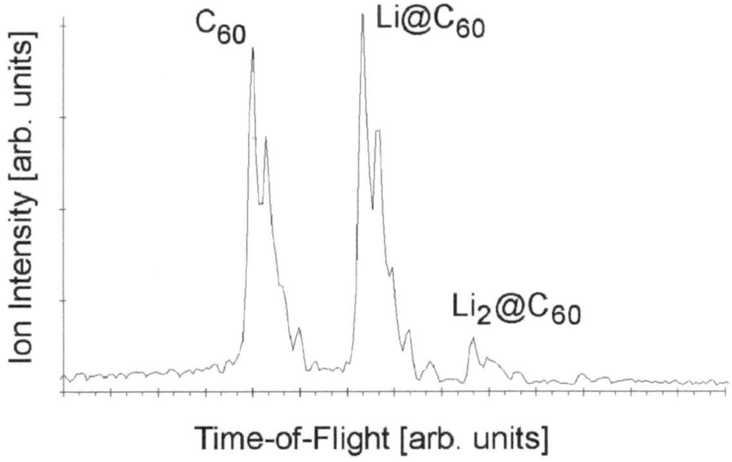

Figure 8.11. Time-of-flight mass spectrum from fullerene film implanted with a high dose of Li^+ ions at an energy of 30 eV.

The yield of endohedral species, as a function of ion implantation energy, as determined by measuring the mass peaks in LDMS mass spectra, is plotted in Fig. 8.12 for the different alkali metals [56]. The absolute yields are not comparable since different ion currents were used for the different alkali metals. The endohedral yield has been compared with the sum of $M@C_{60}^+$ and all endohedral fragment ions measured in the gas phase experiments of Wan et al. (Fig. 8.8 and [49]). The distributions coincide very nicely. In particular, the energetic thresholds are in good agreement as well as the widths of the energetic window within which endohedral fullerene formation is possible. The fact that all endohedral fragments of the gas phase studies have to be taken into account for the comparison reflects the situation that, in the fullerene film, fragmentation is suppressed due to the efficient energy transfer from the collisionally excited fullerenes into the film. No evidence for exohedral MC_{60} production during laser desorption was found for the small alkalis (Li and Na). However, there is a significant exohedral MC_{60} production for K and Rb [56]. This shows up as an ion energy independent background signal on the plots in Fig. 8.12 for energies below 100 eV (beyond this the fullerenes in the film are destroyed on ion impact and the signal decreases). The magnitude of this "exohedral" signal depends on the amount of alkalis present in the film (i.e. on the ion current used for production) and on desorption laser fluence but not on the ion energy.

Superimposed on this background signal is the genuine endohedral signal. As the size of the alkali increases, the energetic threshold for the endohedral production increases and the maximum in the cross section moves to higher energy values just as one would expect from the gas phase results.

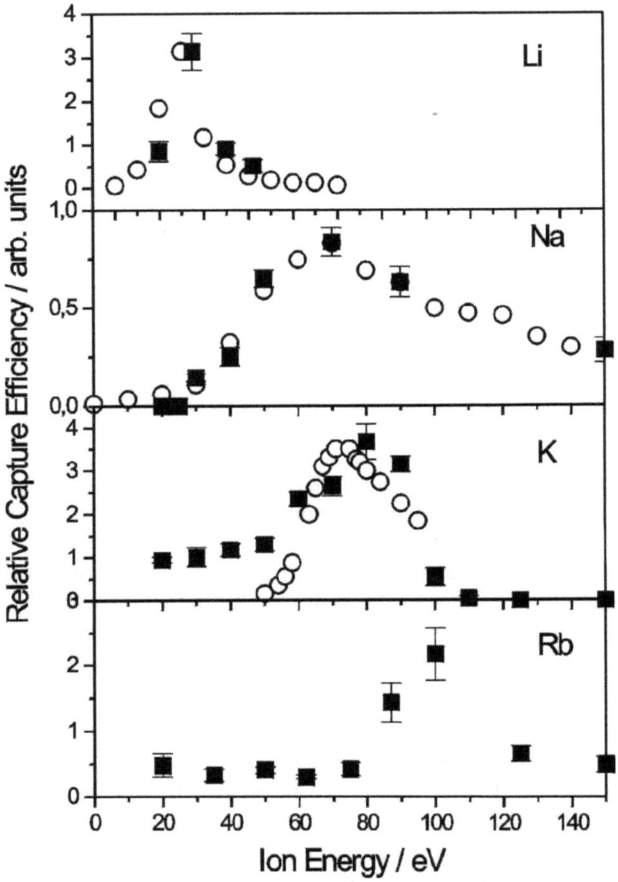

Figure 8.12. Comparison of the endohedral yield of material obtained in alkali ion – fullerene film collisions (squares) [56] and the sum of all endohedral product ions from the gas phase experiments of Anderson and co-workers (circles) [48]. The endohedral fraction of the ion-implanted material was determined by laser desorption mass spectrometry.

Other critical parameters for optimising the efficiency of the production process are the ion current (i.e. the ratio of metal ions to fullerene molecules deposited) and the film thickness. The alkalis show quite different dependencies on the ion current [59]. The differences can be explained in terms of the amount of energy transferred to the fullerene film per ion – the larger the alkali ion, the higher is the ion energy required for endohedral fullerene production and the higher is the likelihood of destroying the fullerenes. However, as discussed briefly above, the films produced with high currents of Li and Na ions are not very soluble and there appears to be a strong degree of polymerisation occurring for high ion currents which does not show up in LDMS.

Studies of the dependency of the endohedral production efficiency on the thickness of the fullerene film deposited on each rotation of the metal cylinder showed that the optimum conditions are for the deposition of one fullerene monolayer per rotation. The capture rate drops by more than a factor of two when going from monolayer to multiple layer deposition per rotation. Finally, there is also an influence of the total thickness of the deposited film. The efficiency decreases practically linearly by almost a factor of two as the total film thickness is increased from a few monolayers to between 50 and 100 nm. After this value the behaviour reaches a plateau, staying constant up to very large film thicknesses [59]. This is a consequence of the efficiency of energy transport within the films and will possibly also depend on the conductivity of the underlying substrate, although this has not been investigated systematically.

It is clear from the above that a good control of the deposition parameters is essential for optimisation of the production method. It is also important to avoid neutral metal vapour from reaching the substrate. If large amounts of neutral metal are present the dominant extractable reaction products are small oligomers (dimers and trimers) of empty fullerenes. This presumably happens in a reaction process closely related to the ball-milling production method for $(C_{60})_2$ [60]. The ion irradiation serves to heat up the fullerene substrate thus facilitating the alkali-metal catalysed oligomerisation reactions. Such reactions do not occur to a noticeable degree when only metal ions are present (at least at the currents routinely used for production).

Reports have also been made of a high efficiency production of endohedral alkali fullerenes using a plasma method [61]. However, the ion energy dependence of the production efficiency is quite different from that discussed here (as determined by LDMS), indicating a different formation mechanism, and purification and further analysis of the purified material is needed to ensure that it is the same product.

Low energy ion implantation can also be used to produce endohedrally doped fullerene derivatives. For example, hydrogenated $C_{60}H_{36}$, prepared by the transfer hydrogenation reaction [62] was deposited at low temperatures and irradiated with 30 eV Li^+. The endohedral fragment ions containing even numbers of hydrogen atoms are more dominant in LDMS spectra and thus more stable than those containing odd numbers [59]. This is the opposite tendency from the empty hydrogenated fullerenes [62] and reflects the transfer of charge from the Li to the fullerene cage. No attempts have yet been made to extract the endohedral hydrogenated fullerenes using high performance liquid chromatography (HPLC).

Recent spectroscopical studies of the purified Li@C_{60} materials obtained from the low energy ion implantation production method have shown that the two fractions that are purified with HPLC are in fact dimers and trimers [63]. However, when the purified samples are evaporated at temperatures of a few hundred °C the weak inter-cage bonds are broken and the properties of monomer molecules can be investigated in the gas phase [64] or they can be deposited on substrates [65].

A similar method was developed by Weidinger and co-workers to implant fullerenes with N^+ and P^+ ions [66, 67]. It has proved to be very difficult to purify and isolate N@C_{60}, however, the molecule shows a very sharp line in EPR spectra that can be used as a fingerprint. Due to the long spin lifetime, it has also been discussed as a possible candidate for quantum computation applications [68].

The studies described in this section provide a nice example of a situation where detailed, fundamental gas phase studies provided the know-how and inspiration for the development of a new macroscopic production method for interesting new nanomaterials.

2.2 Nuclear Recoil

Braun and Rausch [69] showed evidence for radionuclide endohedral fullerene formation from neutron irradiation of polycrystalline C_{60} as a result of Ar impurities trapped in the solid. They proposed that the endohedral nucleotide formation was the result of penetration of radioactive ^{41}Ar atoms into the fullerene cages on account of the high nuclear recoil energy produced from prompt gamma emission following the nuclear reaction ^{40}Ar(n,γ)^{41}Ar. These results were later confirmed by Gadd and co-workers [42]. ^{127}Xe and ^{79}Kr endohedral compounds were also produced using a nuclear recoil method by mixing the fullerene material with K^{127}I or K^{79}Br respectively and irradiating the mixed material with 16 MeV deuterons [39]. Similarly the incorporation of tritium atoms (via neutron activation of ^{6}Li) [70] and ^{7}Be via neutron activation of ^{7}Li and ^{12}C [71] has been shown among many other reactions [72].

The recoil energy of a radionuclide resulting from prompt gamma emission following neutron activation is given by

$$E_{recoil} = \frac{E_\gamma^2}{2m_{recoil}c^2} \tag{8.2.1}$$

where E_{recoil} is the recoil energy of a radionuclide with mass m_{recoil} when a γ-ray of energy E_γ is emitted. The maximum energy available for γ-emission is determined by the Q-value for the neutron activation reaction. This parameter is defined as the difference between the nuclidic masses of the reactants and those of the products and is easily obtained from nuclear data tables [73]. The maximum recoil energies that are obtained for various rare gas nucleotides are on the order of 200-500 eV in the centre-of-mass frame of reference [74]. These energies are rather high compared to those discussed earlier in this chapter but the formation of endohedral fullerenes under these conditions is certainly possible via mechanisms of the kind discussed for the gas phase reactions. The first volume of this book series "Nuclear and Radiation Chemical Approaches to Fullerene Science" [75] contains a number of chapters related to the nuclear recoil approach and the properties of radioactive endohedral fullerenes.

REFERENCES

1 T. Akasaka and S. Nagase, *Endofullerenes. A New Family of Carbon Clusters* (Kluwer Academic Publishers, Dordrecht, 2002).
2 H. Schwarz, Angew. Chem. Int. Ed. Engl. **32**, 1412 (1993).
3 H. W. Kroto, J. R. Heath, S. C. O'Brien, R. F. Curl and R. E. Smalley, Nature **318**, 162 (1985).
4 J. R. Heath, S. C. O'Brien, Q. Zhang, Y. Liu, R. F. Curl, H. W. Kroto, F. K. Tittel, *et al.*, J. Am. Chem. Soc. **107**, 7779 (1985).
5 F. D. Weiss, S. C. O'Brien, J. L. Elkind, R. F. Curl and R. E. Smalley, J. Am. Chem. Soc. **110**, 4464 (1988).
6 Y. Chai, T. Guo, C. M. Jin, R. Haufler, L. P. F. Chibante, J. Fure, L. Wang, *et al.*, J. Phys. Chem. **95**, 7564 (1991).
7 T. Guo, M. D. Diener, Y. Chai, M. J. Alford, R. E. Haufler, S. M. McClure, T. Ohrer, *et al.*, Science ? (1993).
8 S. W. McElvany, M. M. Ross and J. H. Callahan, Acc. Chem. Res. **25**, 162 (1992).
9 H. Shinohara, Rep. Prog. Phys. **63**, 843 (2000).
10 Y. Kubozono, H. Maeda, Y. Takabayashi, K. Hiraoka, T. Nakai, S. Kashino, S. Emura, *et al.*, J. Am. Chem. Soc. **118**, 6998 (1996).
11 L. J. Wilson, D. W. Cagle, T. P. Thrash, S. J. Kennel, S. Mirzadeh, J. M. Alford and G. J. Ehrhardt, Coord. Chem. Rev. **190-192**, 199 (1999).
12 T. Weiske, D. K. Böhme, J. Hrusak, W. Krätschmer and H. Schwarz, Angew. Chem. Int. Ed. Engl. **30**, 884 (1991).
13 J. Hrusák, D. K. Böhme, T. Weiske and H. Schwarz, Chem. Phys. Lett. **193**, 97 (1992).
14 M. Kolb and W. Thiel, J. Comp. Chem. **14 (1)**, 37 (1993).
15 K. A. Caldwell, D. E. Giblin, C. S. Hsu, D. Cox and M. L. Gross, J. Am. Chem. Soc. **113**, 8519 (1991).
16 M. M. Ross and J. H. Callahan, J. Phys. Chem. **95**, 5720 (1991).
17 E. E. B. Campbell, R. Ehlich, A. Hielscher, J. M. A. Frazao and I. V. Hertel, Z. Phys. D **23**, 1 (1992).
18 R. C. Mowrey, M. M. Ross and J. H. Callahan, J. Phys. Chem. **96**, 4755 (1992).

19 T. Weiske, D. K. Böhme and H. Schwarz, J. Phys. Chem. **95** (1991).

20 T. Weiske and H. Schwarz, Angew. Chem. **104**, 639 (1992).

21 T. Weiske, T. Wong, W. Krätschmer, J. K. Terlouw and H. Schwarz, Angew. Chem. Int. Ed. Engl. **31**, 183 (1992).

22 J. Callahan, M. M. Ross, T. Weiske and H. Schwarz, J. Phys. Chem. **97**, 20 (1993).

23 E. E. B. Campbell, R. Ehlich, G. Heusler, O. Knospe and H. Sprang, Chem. Phys. **239**, 299 (1998).

24 R. Kleiser, H. Sprang, S. Furrer and E. E. B. Campbell, Z. Phys. D **28**, 89 (1993).

25 H. Sprang, A. Mahlkow and E. E. B. Campbell, Chem. Phys. Lett. **227**, 91 (1994).

26 J. F. Christian, Z. Wan and S. L. Anderson, J. Chem. Phys. **99**, 3468 (1993).

27 Z. Wan, J. F. Christian and S. L. Anderson, J. Chem. Phys. **96**, 3344 (1992).

28 S. Matejecik, T. D. Märk, P. Spanel, D. Smith, T. Jaffke and E. Illenberger, J. Chem. Phys. **102**, 2516 (1994).

29 R. Ehlich, E. E. B. Campbell, O. Knospe and R. Schmidt, Z. Phys. D **28**, 153 (1993).

30 T. Weiske, T. Wong, W. Krätschmer, J. K. Terlouw and H. Schwarz, Angew. Chem. Int. Ed. Engl. **31**, 183 (1992).

31 R. L. Mury and G. E. Scuseria, Science **263**, 791 (1994).

32 S. G. Kim and D. Tománek, Phys. Rev. Lett. **72**, 2418 (1994).

33 R. L. Murry, D. L. Strout and G. E. Scuseria, Int. J. Mass Spec. Ion Proc. **138**, 113 (1994).

34 J. Laskin, H. A. Jimenez-Vazquez, R. Shimshi, M. Saunders, M. S. d. Vries and C. Lifshitz, Chem. Phys. Lett. **242**, 249 (1995).

35 R. D. Levine and R. B. Bernstein, *Molecular Reaction Dynamics and Chemical Reactivity* (Oxford University Press, New York, Oxford, 1987).

36 K. A. Caldwell, D. E. Giblin and M. L. Gross, J. Am. chem. Soc. **114**, 3743 (1992).

37 F. Z. Cui, D. X. Liao and H. D. Li, Physics Letters A **195**, 156 (1994).

38 R. Ehlich, O. Knospe and R. Schmidt, J. Phys. B **30**, 5429 (1997).

39 T. Ohtsuki, K. Ohno, K. Shiga, Y. Kawazoe, Y. Maruyama and K. Masumoto, Phys. Rev. Lett. **81**, 967 (1998).

40 C. Brink, P. Hvelplund, H. Shen, H. A. Jiménez-Vásquez, R. J. Cross and M. Saunders, Chem. Phys. Lett. **290**, 551 (1998).

41 M. Saunders, H. A. Jimenez-Vazquez, R. J. Cross, S. Mroczkowski, M. L. Gross, D. E. Giblin and R. J. Poreda, J. Am. Chem. Soc. **116**, 2193 (1994).

42 G. E. Gadd, P. J. Evans, D. J. Hurwood, P. L. Morgan, S. Moricca, N. Webb, J. Holmes, *et al.*, Chem. Phys. Lett. **270**, 108 (1997).

43 M. Saunders, H. A. Jiménez-Vázquez, R. J. Cross and R. J. Poreda, Science **259**, 1428 (1993).

44 M. Saunders, H. A. Jiménez-Vázquez, R. J. Cross, S. Mroczkowski, D. I. Freedberg and F. A. L. Anet, Nature **367**, 256 (1994).

45 M. Saunders, R. J. Cross, H. A. Jimenez-Vasquez, R. Shimshi and A. Khong, Science **271**, 1693 (1996).

46 B. A. DiCamillo, R. L. Hettich, G. Guiochon, R. N. Compton, M. Saunders, H. A. Jimenez-Vazquez, A. Khong, *et al.*, J. Phys. Chem. **100**, 9197 (1996).

47 K. Yamamoto, M. Saunders, A. Khong, R. J. Cross, M. Grayson, M. L. Gross, A. F. Benedetto, *et al.*, J. Am. Chem. Soc. **121**, 1591 (1999).

48 Z. Wan, J. F. Christian and S. L. Anderson, Phys. Rev. Lett. **69**, 1352 (1992).

49 Z. Wan, J. F. Christian, Y. Basir and S. L. Anderson, J. Chem. Phys. **99**, 5858 (1993).

50 V. Bernshtein and I. Oref, J. Chem. Phys. **109**, 9811 (1998).

51 K. Ohno, Y. Maruyama, K. Esfarjani, Y. Kawazoe, N. Sato, R. Hatakeyama, T. Hirata, *et al.*, Phys. Rev. Lett. **76**, 3590 (1996).

52 R. Deng and O. Echt, Chem. Phys. Lett. **353**, 11 (2002).

53 S. Tomita, J. U. Andersen, K. Hansen and P. Hvelplund, , submitted (2003).
54 J. Laskin, T. Peres, A. Khong, H. A. Jimenez-Vazquez, R. J. Cross, M. Saunders, D. S. Bethune, *et al.*, Int. J. Mass Spec. **185-187**, 61 (1999).
55 R. Tellgmann, N. Krawez, S.-H. Lin, I. V. Hertel and E. E. B. Campbell, Nature **382**, 407 (1996).
56 E. E. B. Campbell, R. Tellgmann, N. Krawez and I. V. Hertel, J. Phys. and Chem. of Solids **58**, 1763 (1997).
57 B. Winter, R. Mitzner, C. Kusch, E. E. B. Campbell and I. V. Hertel, J. Chem. Phys. **104**, 9179 (1996).
58 A. Lassesson, F. Rohmund and E. E. B. Campbell, Mol. Materials **13**, 225 (2000).
59 E. E. B. Campbell, in *"Endofullerenes: a new family of carbon clusters"*, edited by T. Akasaka and S. Nagase (Kluwer Academic Publishers, Dordrecht, 2002).
60 S. Lebedkin, A. Gromov, S. Giesa, R. Gleiter, B. Renker, H. Rietschel and W. Krätschmer, Chem. Phys. Lett. **285**, 210 (1998).
61 T. Hirata, R. Hatakeyama, T. Mieno and N. Sato, J. Vac. Sci. and Tech. A **14**, 615 (1996).
62 M. Gerst, H.-D. Beckhaus, C. Ruechardt, E. E. B. Campbell and R. Tellgmann, Tetrahedron Lett. **34**, 7729 (1993).
63 A. Gromov, D. Ostrovskii, A. Lassesson, M. Jönsson and E. E. B. Campbell, J. Phys. Chem. B , in press (2003).
64 F. Rohmund, A. V. Bulgakov, M. Heden, A. Lassesson and E. E. B. Campbell, Chem. Phys. Lett. **323**, 173 (2000).
65 V. Popok, A. Lassesson, M. Jönsson, A. Gromov, A. Taninaka, H. Shinohara and E. E. B. Campbell, in *Electrochemical Society, Fullerene and Nanotube Symposium* (Electrochemical Soc., paris, 2003).
66 T. Almeida-Murphy, T. Pawlik, A. Weidinger, M. Höhne, R. Alcala and J.-M. Spaeth, Phys. Rev. Lett. **77**, 1075 (1996).
67 B. Pietzak, A. Weidinger, K.-P. Dinse and A. Hirsch, in *"Endofullerenes: a new family of carbon clusters"*, edited by T. Akasaka and S. Nagase (Kluwer Academic Publishers, Dordrecht, 2002).
68 W. Harneit, M. Waiblinger, K. Lips, S. Makarov and A. Weidinger, in *AIP Conference Proc.*, 2000), Vol. 544, p. 207.
69 T. Braun and H. Rausch, Chem. Phys. Lett. **237**, 443 (1995).
70 H. A. Jiménez-Vázquez, R. J. Cross, M. Saunders and R. J. Poreda, Chem. Phys. Lett. **229**, 111 (1994).
71 T. Ohtsuki, K. Masumoto, K. Ohno, Y. Maruyama, Y. Kawazoe, K. Sueki and K. Kikuchi, Phys. Rev. Lett. **77**, 3522 (1996).
72 T. Ohtsuki and K. Ohno, in *Nuclear and Radiation Chemical Approaches to Fullerene Science*, edited by T. Braun (Kluwer Academic Publishers, Dordrecht, 2002).
73 E. Browne, R. B. Firestone and V. S. Shirley, (Wiley, New York, 1986).
74 G. E. Gadd, in *Nuclear and Radiation Chemical Approaches to Fullerene Science*, edited by T. Braun (Kluwer Academic Publishers, Dordrecht, 2002).
75 T. Braun, *Nuclear and Radiation Chemical Approaches to Fullerene Science* (Kluwer Academic Publishers, Dordrecht, 2000).

Chapter 9

FULLERENE-FULLERENE COLLISIONS

Collisions between fullerenes provide a means to study cluster-cluster interactions both theoretically and experimentally. This makes it possible to explore the similarities and differences in the dynamical processes in collision systems covering many orders of magnitude in size and collision energy; from heavy ion collisions in nuclear physics to macroscopic liquid droplets or even colliding galaxies. Atomic clusters provide an interesting intermediate-size system that can in principle be explored in considerable detail, both theoretically and experimentally, with respect to collision partner size, binding energy and collision energy. Theoretical investigations of cluster-cluster collisions were pioneered by Schmidt and co-workers for the case of collisions between two small alkali clusters. Using molecular dynamics simulations combined with the density functional formalism (Chap. 3.3.2) they were able to show that a number of reaction channels, analogous to the dynamics of heavy ion collisions and colliding liquid droplets were likely to appear in atomic cluster-cluster collisions, such as the fusion of two clusters [1]. Unfortunately, controlled collisions between mass-selected atomic clusters under single-collision conditions are extremely difficult to study experimentally and so far such studies have only been possible for colliding carbon clusters, or fullerenes. Fullerene-fullerene collisions can be carried out with relatively little experimental effort due to the fact that molecular beams of mass-selected carbon clusters can be produced by simply evaporating purified fullerene material. In this Chapter we discuss the experimental and theoretical work involving the study of molecular fusion and subsequent fragmentation in fullerene-fullerene collisions. We also discuss the other reaction channels that can be observed from these collisions such as scattering, fragmentation without fusion and charge transfer, Fig. 9.1.

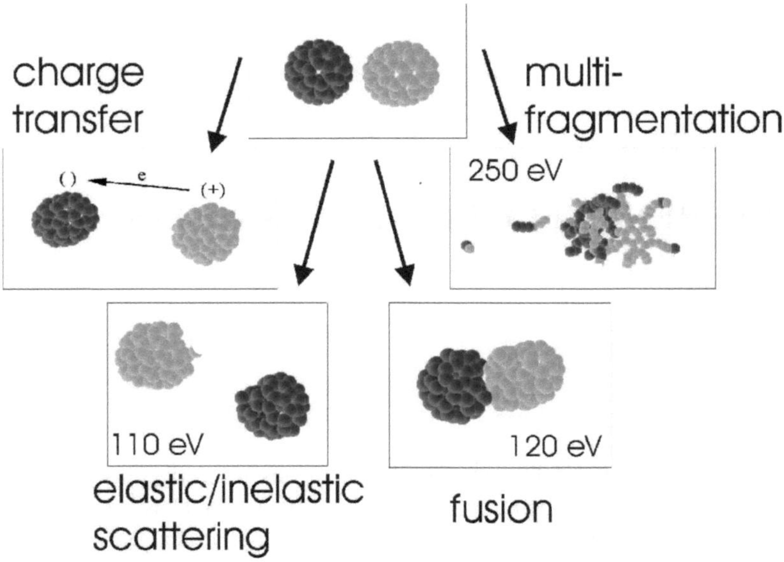

Figure 9.1. Snapshots from MD-DFT simulations of $C_{70}^+ + C_{60}$ collisions illustrating the different observable reactions channels in fullerene-fullerene collisions.

Apart from the very fundamental collision dynamics aspects, fullerene-fullerene collisions are related to other areas of fullerene science and technology where aggregation of fullerene molecules plays a role, e.g. aggregations in the laser plume during laser desorption [2,3], in fullerene-fullerite surface collisions (Chap. 10) and during polymerisation of fullerenes [4]. In addition, insight into the growth of larger fullerenes may be obtained from fullerene-fullerene fusion studies. The experiments and theoretical studies described in detail in this chapter are carried out under well-defined (single collision) conditions in order to provide information on the relevant dynamical processes in terms of reaction cross sections. This information can then be applied in order to improve the understanding of the above-mentioned processes which generally occur under less well-defined conditions.

Before going into the detailed discussion of the topic, it is useful to get a general picture of possible reaction channels in fullerene-fullerene collisions and of the collision energies of interest. Figure 9.1 shows some snapshots from MD-DFT calculations (Chap. 3.3.2) illustrating the reaction products that can be expected in collisions between a singly charged C_{70}^+ and a neutral C_{60} at different centre-of-mass energies. For low collision energies the main reaction channels are charge transfer and inelastic scattering. For low impact parameter collisions, the two fullerenes are compressed and deformed on collision and "bounce" apart again with high vibrational excitation. Molecular fusion is a significant reaction channel for collision energies of around 100-200 eV where a metastable fullerene-like product is formed. As the collision energy increases beyond this, the increasing amount of vibrational excitation in the fusion product leads to increasing sequential evaporation of particles, even on the ps time scale of the simulations. At energies > 200 eV the reaction product fragments into multiple fragments on the 100 fs to 1 ps time scale.

By combining experimental results with simple phenomenological models, statistical modelling, classical MD and detailed but time-consuming MD-DFT simulations it is possible to obtain considerable insight into the dynamics of these collisions. This is the topic of the following section. All collision energies given will be in the centre-of-mass reference frame, unless otherwise stated.

1. FUSION AND FRAGMENTATION

1.1 Fusion

The first observation of molecular fusion between fullerenes was reported in 1993 [5] and was the starting point for a series of experiments on the reaction dynamics in low energy (50eV - 6keV) fullerene-fullerene collisions [6-11]. The experimental work was accompanied by molecular dynamics simulation studies by the Schmidt group and others [12-20]. The experiments were performed in the apparatus shown in Fig. 2.8. The positively charged products of the collision reaction were detected by a time-of-flight reflectron mass spectrometer. The reflectron could be rotated around the scattering cell allowing the determination of the angular distribution of ions. This was not considered in the early experiments [5-7] and in these papers the cross sections for fusion were underestimated at high energies where substantial fragmentation of the fusion product occurs. Newer measurements determined the angular distributions of the fusion products and integrated over these to obtain corrected values of the absolute cross sections [8,21]. A typical series of "raw" mass spectra is shown in Fig. 9.2 for a scattering angle of $0 \pm 0.5^{\circ}$. The energy window in which a clear signal of a fusion product can be seen is rather small. It is possible to distinguish between fused reaction products and inelastically scattered projectile ions due to the kinematics of the collisions. When two fullerenes fuse together the velocity of the products and any fragments from a fused product are around half of the velocity of the parent ion, Fig. 9.3. This

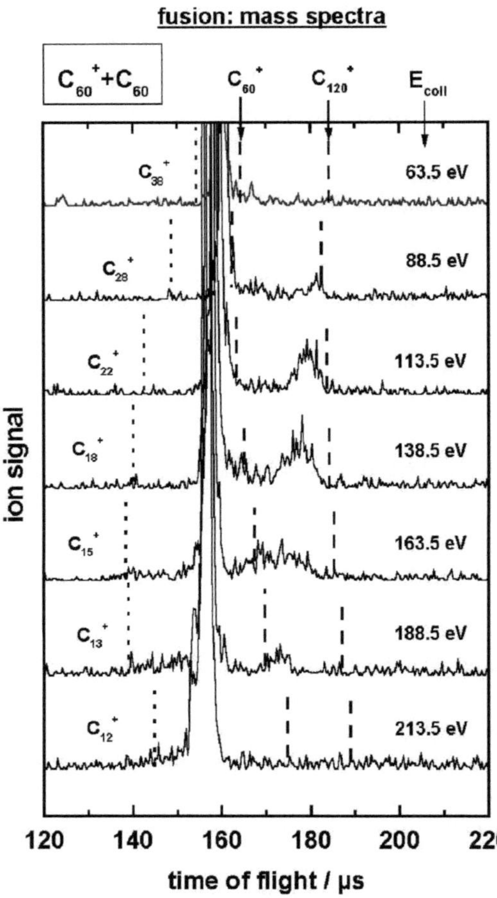

Figure 9.2. Reflectron time-of-flight spectra of $C_{60}^+ + C_{60}$ collisions at different centre-of-mass collision energies. At the lowest and highest collision energy shown there is no clearly discernible signal arising from fused fullerenes. Beyond the fusion threshold energy the product ion distribution broadens and shifts to lower masses as the collision energy is increased. Expected arrival times for the fusion products and some of their fragment ions are marked. Adapted from [7]

allows them to be separated in a reflectron time-of-flight mass spectrometer. However, as can be seen in Fig. 9.2, the arrival times of small fragments from a fusion product overlap with large fragments from scattered projectiles. In order to separate out the different contributions to the time-of-flight spectrum an additional pulsed electric field was used to accelerate and detect the products of the collision. This allowed all fusion products to be clearly separated from non-fused projectile ion fragments [9].

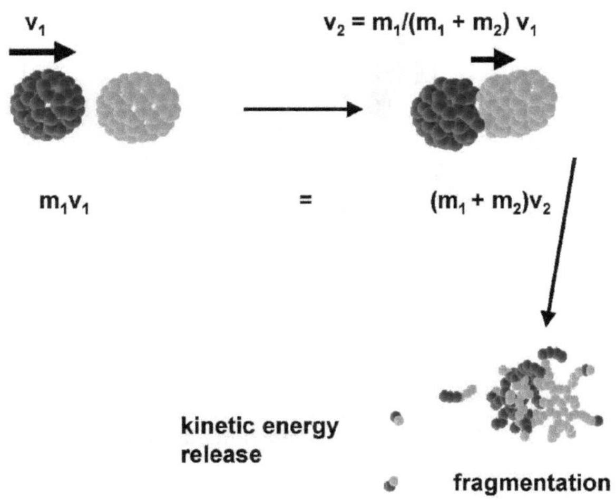

Figure 9.3. The velocity of the products of a fusion reaction between C_{60}^+ and C_{60} is half that of the projectile ion plus the influence from kinetic energy release on fragmentation.

The fusion cross sections are obtained from integrating over the intensities of all product ions with a laboratory velocity corresponding to that of fusion products, after corrections to the intensities have been made due to the detection probability at the micro-channel plate ion detector [7] (see Chap. 2.1.3). The absolute fusion cross section for $C_{60}^+ + C_{60}$ collisions is plotted in Figure 9.4. Note that the error bars shown are relative errors in the measurement. There is a large error with respect to the absolute cross section due mainly to uncertainty in the fullerene vapour pressure.

It is tempting to compare the form of the cluster fusion cross sections with those obtained in nuclear collisions. Fig. 9.4(b) gives a schematic picture of the behaviour in such collisions. The rise in the cross section at energies close to threshold can be well described using a simple line-of-centres or absorbing sphere model [22]. The fall in the cross section towards higher energies is due to the centrifugal instability of the fused compound. Fusion is an activated process and the simplest expectation is that the magnitude and energy dependence of the fusion cross section will be limited

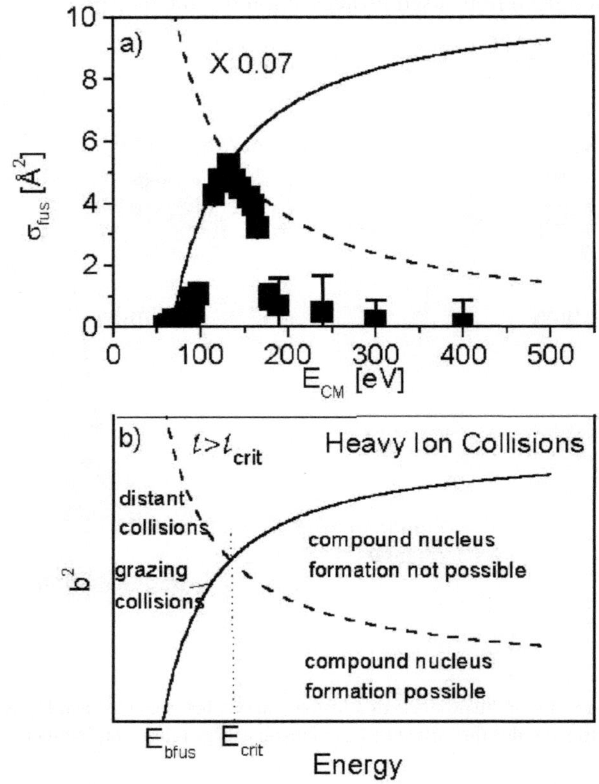

Figure 9.4. (a) Measured absolute fusion cross section for $C_{60}^+ + C_{60}$ collisions. The experimental results are compared with a simple line-of-centres (or absorbing sphere) model which is often used to describe the fusion reaction in nuclear heavy ion collisions. Note that the calculated cross sections have been scaled by a factor of 0.07.
(b) schematic cross section behaviour for nuclear collisions.

by the condition that the energy available along the line of centres of the two fullerenes (or colliding nuclei) exceeds this threshold. At an initial relative translational energy E and impact parameter b the energy along the line of centres is $E = (1\text{-}b^2/R^2)$ where R is the centre-to-centre distance. We locate the barrier to fusion at a distance R equal to the sum of the radii of the collision partners $R = d \approx 7$Å for two C_{60}. Then the maximum impact parameter that allows the crossing of the barrier is determined by $E = (1\text{-}b_{max}^2/R^2) = E_{bfus}$ or

$$\sigma_{fus} = \begin{cases} 0, & E \le E_{bfus} \\ \pi d^2 \left(1 - E_{bfus} / E\right), & E > E_{bfus} \end{cases} \tag{9.1.1}$$

By plotting the fusion cross section as a function of *1/E* it is possible to determine the energetic threshold for the reaction, assuming the collision can be well described in this way. However, the measured fullerene fusion cross section actually differs from the simple model in several key aspects. The rise in threshold is not quite linear in *1/E*, Fig. 9.5, and the slope is quite wrong. The geometrical forefactor is much smaller than πd^2. It is well known that the line-of-centres model typically overestimates the magnitude of the cross sections for chemical reactions. This is often ascribed to steric requirements of the reactions and the simplest correction is to lower the probability of reaction by a "steric factor" p, $p<1$ [6,7,23]. The forefactor is then $\pi d^2 p$ and fitting values of p to the experimental results (after correcting for the scattering losses, not considered in the original articles) yields values from 0.07 for C_{60}^{+}+C_{60} to 0.4 for C_{70}^{+}+C_{60} or C_{60}^{+}+C_{70} and 0.45 for $C_{/0}^{+}$+C_{70} collisions. A more refined approach makes the steric factor also depend upon the impact parameter but to compute it requires knowledge of the barrier as a function of the approach geometry. The experimental threshold energies for the various fullerene fusion reactions were determined by assuming that the steric factor p increases linearly with collision energy. This gives a functional form

$$\sigma_{fus} \propto \frac{\left(E - E_{bfus}\right)^2}{E} \tag{9.1.2}$$

which provides a very good fit to the experimental data [7]. The experimental fusion threshold energies, determined by fitting eq. (4.2) to the data, are tabulated in Table 9.1. Such a dependence was rationalised by assuming that different orientations of the fullerenes on impact have different energetic barriers to fusion.

Various molecular dynamics studies have been applied to determining the threshold energy for fullerene fusion [14,20,23,24]. The results of these studies are also given in Table 9.1. Note that there are also rather large errors in the MD determinations of the threshold energy due to the limited number of trajectories calculated but these have not been given explicitly. The simulations all give rather good agreement with experiment as long as the initial internal energy of the projectile ions is properly accounted for. However, extensive MD-DFT simulations [11,23] showed no evidence for a dependence of the fusion probability on the relative orientation of the colliding cages. The dependence on the initial orientation was tested in two

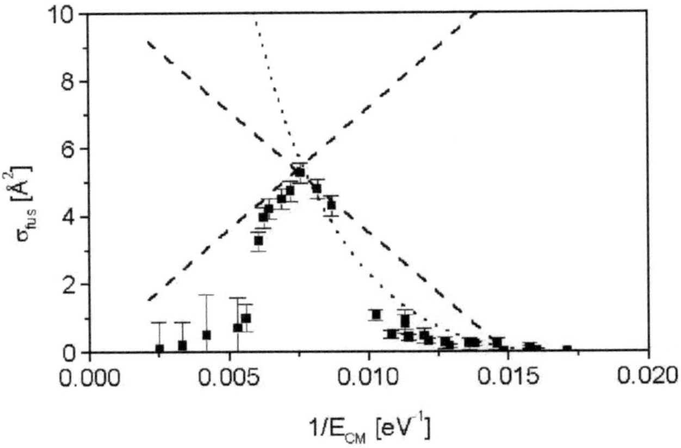

Figure 9.5. Absolute cross section for fusion in C_{60}^{+} + C_{60} collisions plotted as 1/E. The threshold energy for the reaction is determined by fitting eq (9.1.2) to the experimental data, dotted line. The simple line-of-centers model (with cut-off due to centrifugal instability), dashed line (linear dependence on 1/E close to threshold) does not give a good fit to the data. From [21].

different ways. First, two fullerenes were oriented hexagon-to-hexagon (6x6) or pentagon to pentagon (5x5). Then one of the fullerenes was rotated around the collision axis by the angle $\delta\theta$. The collision trajectory was calculated for each value of $\delta\theta$ and for several collision energies [8,11]. The probability for fusion was found to be more dependent on the collision energy close to threshold than on the initial orientation. Results for C_{60}^{+} + C_{60} collisions calculated with MD-DFT are given in Table 2. No correlation with orientation can be found but the number of different orientations that lead to fusion does increase with increasing collision energy. Secondly the principal orientation of C_{70} was chosen prior to C_{70} + C_{60} collisions [23]. The simulations showed no dependence on the orientation of the symmetry axis of C_{70}. However, the absolute cross sections calculated with MD-DFT are in excellent agreement with the experimental data in contrast to those calculated with classical MD [8]. The absolute cross sections for the different collision systems studied are shown in Fig. 9.6 along with the results of the MD simulations. Unfortunately, not enough trajectories have been calculated for C_{60}^{+} + C_{60} using the MD-DFT approach to give a reliable estimate of the fusion cross section. An early report gave an average calculated fusion probability close to threshold of 0.06 which is in agreement with the experimental values [6]. Although classical MD gives a reasonable value for the threshold energy it vastly overestimates the absolute magnitude of the cross sections and also the width of the collision energy window for which fusion is an observable reaction channel.

Table 9-1. Threshold energies for molecular fusion between fullerenes in units of eV

System	Exp. Threshold [7]	Classical MD [8] T = 2000 K	Classical MD [14] T = 2000 K	MD-TB [20] T = 2000K	MD-DFT [8,23] T=2000 K
$C_{60}+C_{60}$	60±1	60	50	60	60
$C_{70}+C_{60}$	70±7	70			70
$C_{70}+C_{70}$	76±4	80			75

Table 9-2. Some results of 6x6 orientation studies for C_{60}^{+} + C_{60} collisions using MD-DFT. One of the fullerenes was rotated about the collision axis by an angle $\delta\theta$. Impact parameter b=0. Here "f" stands for fusion, "s" for scattering without fusion. From [11].

$\delta\theta$	E=110 eV	E=120 eV	E=130 eV	E=140 eV	E=150 eV
0.0	f	s	s	s	f
0.5	s	s	s	f	f
1.0	s	s	s	s	f
1.5	s	s	f	f	f
2.0	s	s	f	f	f
2.5	s	s	s	f	s
3.0	f	s	f	f	f
3.5	s	s	f	s	f
4.0	s	f	f	f	f

Beyond the threshold, the cross section rises rather sharply to a maximum and then falls steeply to higher energies. In nuclear physics this decrease in the probability of fusion is associated with the compound nucleus not being formed. The impact parameter becomes so high that the effective potential $V(R)+Eb^2/R^2$ no longer supports a well. MD simulations show that the product of the collisions close to the fusion threshold takes the form of a peanut, Fig. 9.7. The estimated stability of the peanut shaped isomer of C_{120}^{+} is about that of the reactants [25]. So C_{120}^{+} has a barrier for dissociation which about the same as the fusion barrier. To fill-in the well to the left of the barrier the centrifugal energy must balance the dissociation barrier which gives the condition for the impact parameter

$$b^2 \geq \frac{E_{bfus} d^2}{E} \tag{9.1.3}$$

The energy dependence of the cross section due to such a cut-off is then

$$\sigma_{\text{fus}} = \pi R^2 (E_{\text{bfus}} / E) \tag{9.1.4}$$

Here too, the forefactor is much too large and the decrease as 1/energy is too moderate to fit the steep drop of the experimental cross section (Fig. 9.5).

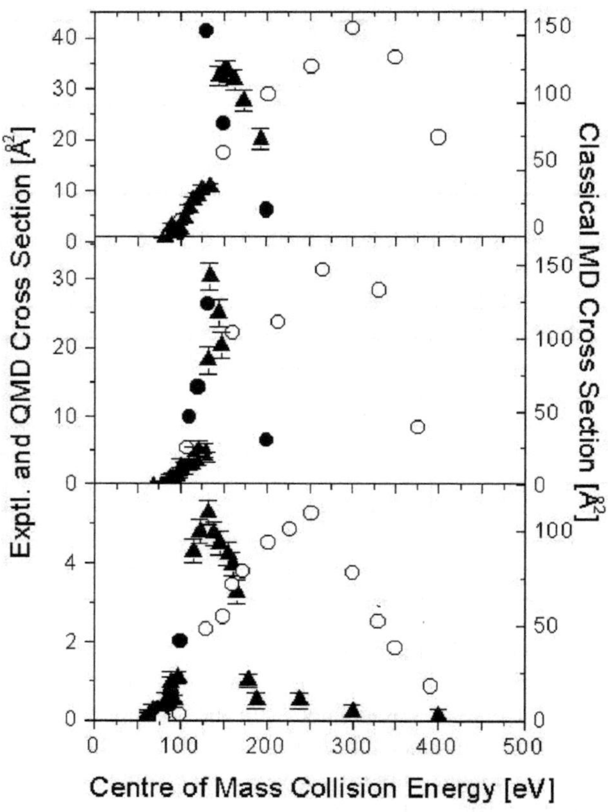

Figure 9.6. Cross sections for fusion between two fullerenes (a) $C_{60}^+ + C_{60}$; (b) $C_{60}^+ + C_{70}$ and $C_{70}^+ + C_{60}$; (c) $C_{70}^+ + C_{70}$. Triangles: experimental data; filled circles: MD-DFT simulations; open circles (right hand axis): classical MD simulations. Adapted from [8].

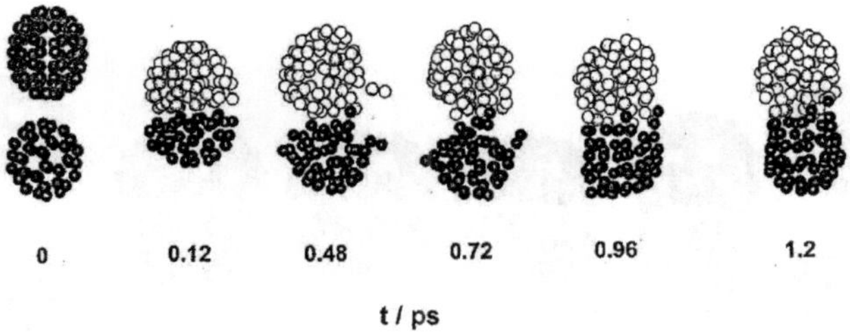

Figure 9.7. MD-DFT simulation of a collision between $C_{70}^{+} + C_{60}$ for a collision energy of 130 eV and b=0. A metastable peanut isomer is formed on a time scale of 1 ps. From [26]

In order to at least partially explain the discrepancies with the simple model one has to consider the possibility of competing reaction channels [21]. In the threshold regime the available phase space for the inelastic scattering channel is larger than that for the formation of a very hot compound fullerene, say C_{120}^{+}. Such a competing channel can be observed in molecular dynamics simulations, Fig. 9.8, and is also seen experimentally as will be discussed in more detail in the following section. To estimate the branching ratio for fullerene-fullerene collisions we require the density of vibrational states of a polyatomic molecule. For our purpose a simple RRK estimate is sufficient because we only need ratios and not absolute numbers. For an *n*-atomic species with *s* vibrational modes, $s = 3N - 6$, and a (geometric) mean vibrational frequency $\overline{\nu}$ (2.7×10^{13} Hz = 900 cm^{-1} for C_{60}) the number of vibrational states below the energy E is

$$N(E) = \frac{1}{s!}\left(\frac{E}{h\overline{\nu}}\right)^{s} \tag{9.1.5}$$

The essential difference between the two channels is that C_{120} has many more vibrational modes than does C_{60}. Once the energy is high enough to reach the vibrational quasi-continuum of C_{120}, it will have overwhelmingly more states than C_{60}. But it takes less energy to reach the quasi-continuum of C_{60}. At the same internal energy, the ratio of the number of states is given by

$$\frac{N(C_{60})}{N(C_{120})} \approx \frac{\left(\dfrac{eE}{sh\overline{\nu}}\right)^{s}}{\left(\dfrac{eE}{2(s+3)h\overline{\nu}}\right)^{2(s+3)}}, \quad s = 174 \tag{9.1.6}$$

Figure 9.8. MD-DFT trajectory showing inelastic scattering where one of the fullerenes emerges from the collision with a high vibrational excitation. The time scale of the simulation is 1 ps. From [11]

The mean energy per vibrational mode needs to exceed the mean vibrational spacing for C_{120} to offer more vibrational states to accommodate the internal energy. In the immediate post threshold region we thus expect that many collisions do not end up in fusion but in inelastic collisions. At higher energies the situation is reversed and, because of the steep rise of the density of states with energy, once formation of C_{120} is the dominant channel, it is the overwhelmingly dominant one. This very simple consideration leads to a much steeper rise of the cross section close to threshold [21], assuming an energy transfer of $0.3E$ in the collisions leading to scattering rather than fragmentation (see 9.1.2 for a justification of this value). The increasing phase space that becomes available for fusion at energies past the barrier means that the fusion cross section rises more rapidly than as $1\text{-}(E_{bfus}/E)$, as seen in figures 9.4, 9.5 and 9.6. However, the experimental situation is rather more complicated with a broad but unknown internal energy distribution for the projectile ions and also a distribution in the energy transferred in the scattering collisions. It is only possible to give an upper limit to the internal energy of the projectiles by considering the amount of metastable fragmentation in the beam. If one assumes that the fusion barrier shifts with the initial internal energy of the projectile and target, as indicated by the results of MD simulations, then it is also possible to obtain an overall cross section with positive curvature but for reasonable assumptions the rise in the cross section at threshold is not as steep as that observed in the experiment.

Fusion is operationally defined as formation of a compound species accompanied by an extensive dissipation of the internal energy. It is however possible for the hot compound species to fragment promptly. This process is very clearly seen in molecular dynamics simulations where many trajectories leading to C_{120} formation also show it breaking apart on a time scale of one or two vibrations. For such short times energy is not fully dissipated and these events are not detected in the experiments as 'fusion'. The competition with the prompt fragmentation (with incomplete energy equilibration) means that the fusion cross section will decrease strongly with energy, past the threshold for three-particle production. What we therefore require is an estimate for the cross section for direct (i.e. prompt) collision induced dissociation (CID). The CID cross section can be computed if all the excess energy is disposed of as kinetic energy of the three outgoing particles, The result is [21]

$$\sigma_{\mathrm{CID}} = A\frac{(E - E_{\mathrm{CID}})^{2.5}}{E} \tag{9.1.7}$$

where A is a geometrical factor and E_{CID} is the threshold for CID via fusion. The branching ratio for fusion is $(\sigma_{\mathrm{com}} - \sigma_{\mathrm{CID}})/\sigma_{\mathrm{com}}$ where σ_{com} is the cross section for complex (=compound species) formation, given by the capture cross section of equation (9.1.1). The fusion cross section is

$$\sigma_{\mathrm{fus}} = \sigma_{\text{2-body}}\left((\sigma_{\mathrm{com}} - \sigma_{\mathrm{CID}})/\sigma_{\mathrm{com}}\right) \tag{9.1.8}$$

where $\sigma_{\text{2-body}}$ incorporates the centrifugal cut-off, equation (9.1.4) and accounts for the competition with scattering close to threshold, as discussed above. The rapid decline of the fusion cross section beyond E_{CID} as given by equation (9.1.7) is in good agreement with the experimental results. This is shown in Fig. 9.9 for $C_{70}^+ + C_{70}$ collisions (the calculations do not take the range of initial internal energies into account which would presumably give a more "gentle" behaviour at the wings). The model works equally well for $C_{60}^+ + C_{70}$ and $C_{70}^+ + C_{60}$ collisions but still overestimates the $C_{60}^+ + C_{60}$ cross section by a factor of 6 although the form of the cross section is correctly reproduced. The reason for the exceptionally small cross section in $C_{60}^+ + C_{60}$ collisions is still not fully understood.

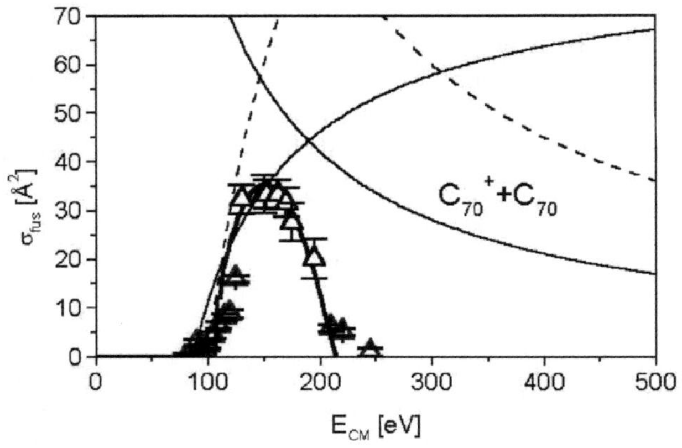

Figure 9.9. Comparison of experimental cross section for C_{70}^{+} + C_{70} collisions with fusion models. Triangles: experimental data. Dashed line: line-of-centres model with centrfugal cut-off. Thin full line: line-of-centres model scaled with $p = 0.45$. Thick full line: fusion cross section with competing scattering and direct collision induced dissociation channels. Adapted from [21].

The experimental and theoretical studies discussed in this sub-chapter were carried out under single-collision conditions. Coalescence and fusion of fullerenes also occurs under less well-defined conditions. Irradiation of fullerene films with laser radiation leads to the formation of fullerene aggregates up to the pentamer [3] with maxima in the corresponding mass spectra at 118, 178, 236 and 300 carbon mass units. These aggregates are very stable, surviving surface reflection at 160 eV collision energy. It can therefore be speculated that they have structures similar to those molecules formed in the collision experiments. It was also shown that in laser desorption, fusion occurs predominantly between C_{60} and a C_{58} fragment. C_{58} is chemically more reactive [27] but also less stable than C_{60}. Thus, the fusion barrier in C_{58}-C_{60} collisions is expected to be lower than in C_{60}-C_{60} collisions. The relative collision energy in the laser desorption experiments is much less than in the single-collision experiments. However, the interaction time is longer and the *total* available energy (kinetic plus internal) is similar in both cases. Molecular fusion of fullerenes was also observed in collisions between C_{60}^{+} ions and fullerene films at 275 eV impact energy [28]. It is very likely that the reaction mechanism in these experiments is similar to the gas phase collisions. High mass products of a fusion reaction at this relatively high collision energy are possible since the substrate can absorb energy, thus stabilising the fused compound and reducing fragmentation.

1.2 Fragmentation

At collision energies slightly above threshold, the fusion product can fragment on the microsecond time scale of the experiment [9]. The average fragment ion mass and the range of fragment ions produced from the fusion product of C_{60}^{+}+C_{60} collisions is plotted in Fig. 9.10 as a function of internal energy (centre of mass collision energy plus initial maximum internal energy). The maximum entropy procedure (Chap. 3.2.2) has been used to characterise the outcome of the collision at these higher energies [21]. These products are classified as coming from fusion because they result from a process where the initial centre of mass energy has been converted to internal energy. The application of the maximal entropy model to the products of collisional fusion is much more straightforward than for laser or collisional excitation of C_{60} [29] since we have a much better knowledge of the energy transferred to internal degrees of freedom in the collision. The same values of binding energies and ionisation potentials for all possible fragments up to and including C_{60}^{+} are taken to be the same as those used earlier for applications of the model to C_{60} and C_{60}^{+} fragmentation [29]. Extrapolation to larger clusters was done by using the following expressions for the binding energy (*BE(n)*) and ionisation potential (*IP(n)*) of cluster C_n in units of eV

$$\begin{aligned} BE(n) &= -7.30n + 20.241 \\ IP(n) &= 5.6 + 13.8/\sqrt{n} \end{aligned} \tag{9.1.9}$$

These expressions are based on [30] but the binding energy of the larger clusters has been reduced slightly to obtain a smoother dependence of binding energy on n on going from $n < 60$ to $n > 60$.

The comparison of the experimentally determined average fragment mass as a function of total internal energy (assuming a total initial internal energy of 20 eV for the projectile ion and 5 ev for the neutral target) [9] with the maximal entropy model is shown in Fig. 9.10 and is rather good. The range of detected fragment ions is shown in Fig. 9.11. The detection limit for fragments predicted by the calculations was taken to be 0.001 (i.e. if the predicted fragment intensity was less than 10^{-3} of the initial parent ion intensity it was not counted). The agreement here is not quite so good. The behaviour at low collision energies follows the experiments very well and the onset of the bimodal fragment distribution is in good agreement. The

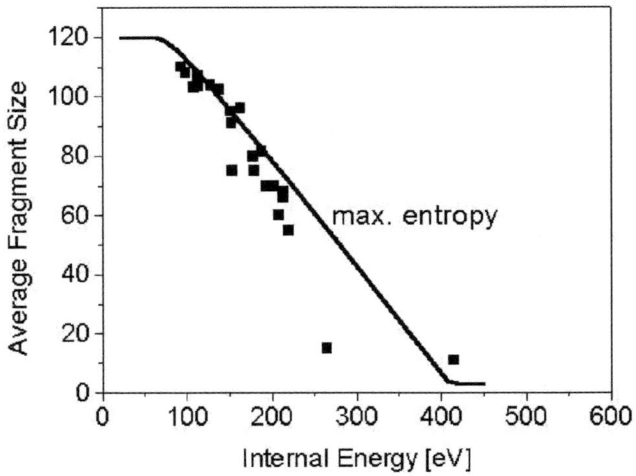

Figure 9.10. Average fragment size from the fusion product in C_{60}^{+} + C_{60} collisions versus total internal excitation energy (centre-of-mass collision energy plus estimated initial internal energy of projectile and target). Squares: experimental data. Full lines: result of maximal entropy calculation. Adapted from [21].

main discrepancy is that the model shows the presence of C_{120}^{+} up to high collision energies beyond the onset of the bimodal fragment distribution. This is not seen in the experiments where there is a steady decrease in the size of the fullerene-like fragments up to a total internal energy of ca. 200 eV. This is not too surprising since the binding energies used in the calculations are those of the most stable fullerene isomers. The agreement can be improved slightly by making some additional assumptions concerning the binding energies of the isomers formed in the collision experiments [21].

Overall the agreement between experiment and the simple statistical model is very good. This provides us with a confirmation that there is equilibration of the centre-of-mass collision energy among the vibrational degrees of freedom of the product of the fusion reactions and that the subsequent fragmentation behaviour is completely statistical in nature. MD-DFT calculations have also provided evidence for the efficient energy equilibration on a time scale of ps [23].

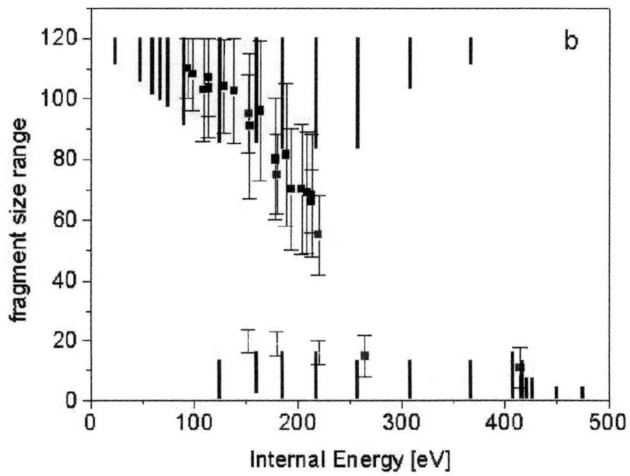

Figure 9.11. Fragment size distribution as a function of total internal energy. Squares: experimental data, thin lines indicate range of detected fragment sizes (not error bars). Thick lines: maximal entropy calculations. Adapted from [21].

The picture of a purely statistical fragmentation mechanism is borne out by studies of the angular dependence of the fragmentation products [9]. Fig. 9.12 shows angular distributions measured for the products of a fusion reaction at three different collision energies. The signal has been integrated over all fragment masses. The distributions are concentrated at small laboratory scattering angles. Assuming that the fused compound formed in the collision undergoes a very rapid energy equilibration and then undergoes a completely statistical fragmentation it is possible to estimate the expected cut-off angle as a function of collision energy. A Newton diagram for the collisions is shown in Fig. 9.13. Immediately after collision, the velocity of the fused compound in the centre-of-mass reference frame will be zero. The final centre-of-mass velocity of the fragment ions (w'_{KER}) will be given by the accumulated kinetic energy released in the fragmentation process. This will be randomly distributed in space. The maximum detectable laboratory scattering angle (θ_{max}) is thus determined by the magnitude of the kinetic energy release which, in turn, depends on the temperature of the fused compound and the mass of the fragment. It was assumed that the small fragment ions ($n < 30$) are produced by a multifragmentation process in which the highly excited fused compound disintegrates into a number of ring and chain fragments, one of which carries the positive charge and is detected in the experiment.

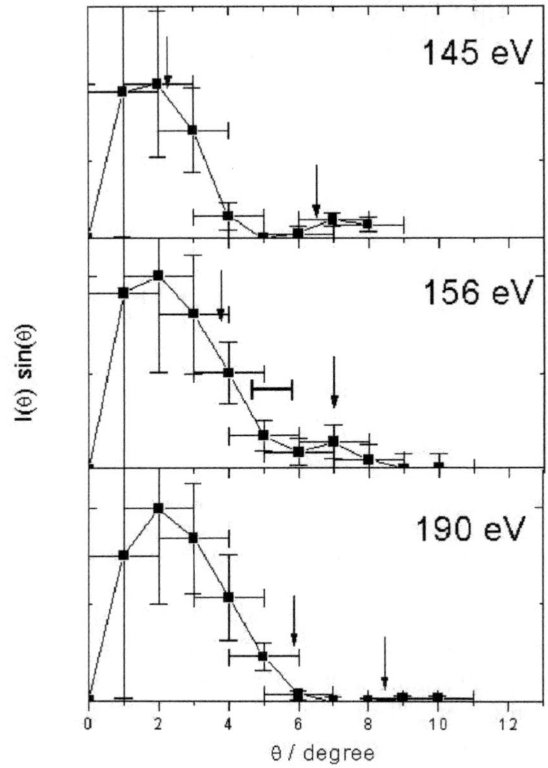

Figure 9.12. Integrated fusion signal intensity in $C_{60}^+ + C_{60}$ collisions, ($I(\theta)sin(\theta)$), plotted as a function of scattering angle. Collision energies of (a) 122 eV (b) 131 eV (c) 165 eV. The energies shown on the plot are the total internal energy (centre of mass collision energy plus initial internal energy of projectile ion and target) Intensities have been normalised to the same maximum value. The arrows indicate the estimated maximum scattering angle for the maximum and minimum sized fullerene-like fragments detected in the experiment. The horizontal bar indicates the estimated maximum scattering angle range for the small fragment ions detected at this collision energy. From [9]

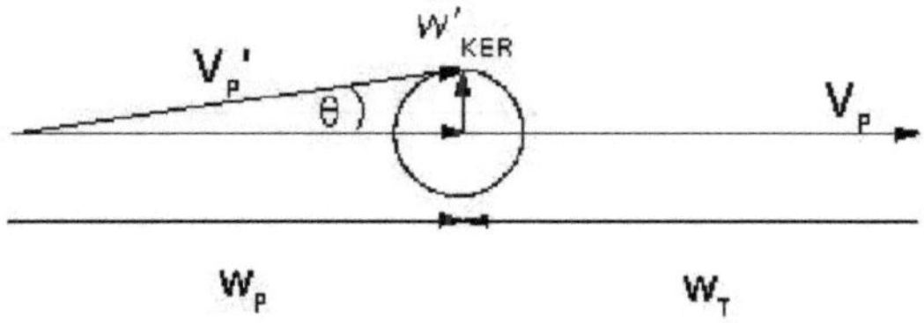

Figure 9.13. Newton diagram illustrating the relationship between laboratory and centre-of-mass frames of reference for fragmentation following fusion of $C_{60}^{+} + C_{60}$.

This may undergo a further evaporative step by emitting a C_3 molecule (the most energetically favourable decay pathway for small carbon clusters). The large fullerene fragment ions are assumed to be produced by successive evaporative loss of C_2 molecules. Obviously this is an over simplification- the small fragments may undergo further evaporative cooling on the time scale of the experiments and the large fragment ions may well be produced via the loss of neutral fragments larger than C_2, especially at high collision energies. However, to a first approximation these seem to be reasonable assumptions. In order to estimate the temperature of the fragments one can make use of the maximal entropy calculations, discussed above. Fig. 9.14 gives the relationship between total internal energy of the fused compound, $<E>$ and the fragment temperature T, as obtained from the maximal entropy model. By treating successive evaporation processes as a random walk, whereby the net velocity after a chain of fragmentation steps is given by $w_{KER} = \sqrt{N}\bar{w}$ with N the number of steps and $\bar{w}$ the average velocity gained by a fragment ion in a single fragmentation, it is possible to estimate the maximum possible laboratory scattering angle from Fig. 9.13. Considering just the maximum and minimum sized fullerene fragment ions detected in the experiments at each collision energy, one can thus estimate the maximum expected laboratory scattering angles. These positions are marked with arrows in Fig. 9.12. The simple estimate gives rather good agreement with the experimental data for the two lowest collision energies. At the highest energy it is likely that the assumption of purely successive C_2 fragmentation breaks down and the maximum laboratory scattering angle is therefore over-estimated. Similar estimates can be carried out for the small fragment ions although they are of low intensity for these collision energies and do not contribute significantly

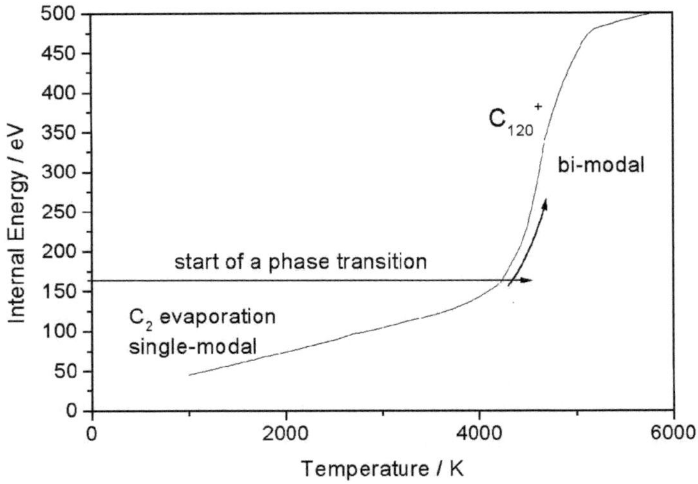

Figure 9.14. Relationship between average internal energy of fused compound, C_{120}^+, and the temperature of its fragments, calculated with the maximal entropy model (Chap. 3.2.2). The onset of the "phase transition" corresponds to the threshold for observation of small ($n > 30$) fragment ions.

to the overall scattering intensity. The horizontal bar on Fig. 9.12b indicates the range of maximum scattering angles at this collision energy for the observed small fragment ion mass range of n = 15-23.

The angular distributions of the integrated fragment ion intensities from scattered projectiles that have not undergone a fusion reaction are shown in Figure 9.15 and are quite different from the distributions from the fusion products in Fig. 9.12. The corresponding average fragment ion sizes are plotted in Fig. 9.16. The angular distribution is centred around a laboratory scattering angle of about 8° for the lowest collision energy investigated of 130 eV. It then broadens and moves towards smaller angles, reaching its maximum width for a centre-of-mass collision energy of 860 eV. As energy is increased beyond this the angular distribution becomes narrower again. Measurements of the ion kinetic energies, using a retarding field energy analyser, show that the fragment ions velocity is about 85% of the initial projectile velocity [9]. With the help of the Newton diagram in Fig. 9.17 it is possible to estimate the amount of energy converted to internal excitation of the projectile and target for projectiles scattered at an angle corresponding to the maximum in the measured distributions. For all collision energies investigated this amounts to close to 60% of the centre-of-mass energy. This is in good agreement with molecular dynamics statistical trajectory ensemble

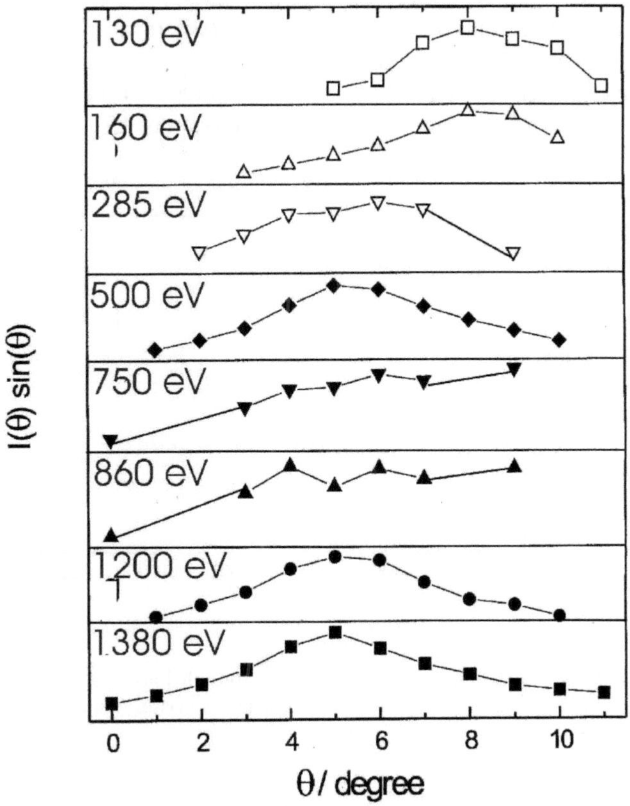

Figure 9.15. Integrated fragment intensities from scattered projectile ions at different collision energies as a function of laboratory scattering angle. Adapted from [9].

simulations at a collision energy of 500 eV that indicated the transfer of ca. 50% of the centre-of-mass energy to internal degrees of freedom [31]. If we further assume that on average this energy is equally divided among the projectile and target and we know the average fragment size from Fig. 9.16 then it is also possible to estimate the width of the angular distribution. We follow the procedure discussed above for the fragments from the fused product. We assume the projectile ions are first scattered inelastically and then subsequently undergo statistical fragmentation. The fragment temperature is obtained from Fig. 3.10. The estimated widths are in very good agreement with the measurements. They increase from ±2° to ±5° on increasing the collision energy from 132 to 500 eV. For collision energies of 750 eV and 860 eV where one still has some fullerene-like fragments as well as small fragment ions (average fragment ion size of around n = 30, Fig. 9.16) the angular distribution is estimated to be ca. ±7°. On further increasing the collision energy beyond this range, the fragment ion mass spectrum consists of only small fragment ions (where we consider only two steps, as discussed above, to estimate the angular distribution - a multifragmentation followed by one evaporative cooling step) and the estimated angular distribution decreases again to about ±5°.

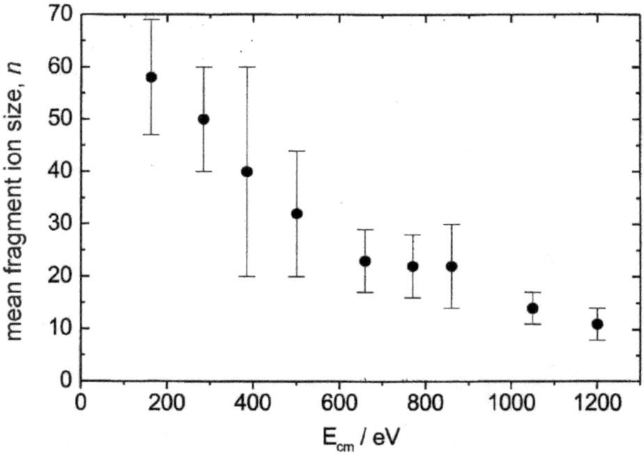

Figure 9.16. Mean fragment ion size from scattered projectile ions as a function of centre of mass collision energy. The bars show the range of detected masses (not error bars). Adapted from [9].

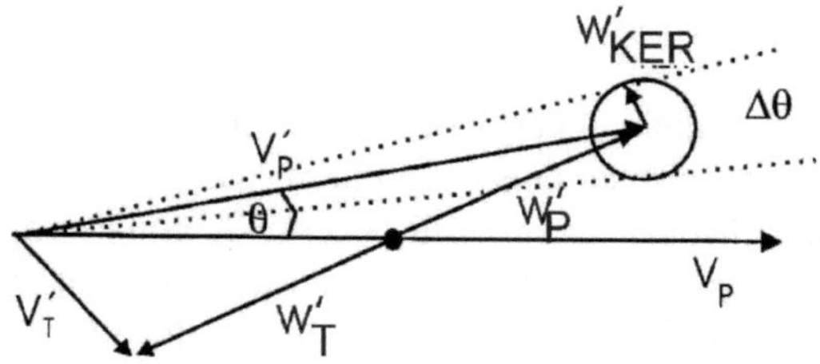

Figure 9.17. Newton diagram for $C_{60}^+ + C_{60}$ collision with inelastic scattering. v: velocity in laboratory system; w: velocity in cente-of-mass system; p: projectile; t: target; w'_{KER}: velocity due to kinetic energy release on fragmentation.

Both the fragmentation of the fused compound and the fragmentation of the inelastically scattered projectile ion are consistent with a completely statistical fragmentation at the low collision energies relevant to these experiments. At the high energy range of the investigated collision energies the fused compound has an extremely short lifetime and breaks up into multiple fragments before the centre-of-mass collision energy can be statistically distributed among all vibrational degrees of freedom. This is the reason for the sharp fall in the fusion cross sections.

The experiments discussed in this section provide convincing evidence for the occurrence of a phase transition and the onset of multifragmentation, even at the relatively low collision energies investigated experimentally. However, the experiments are restricted since only the positively charged fragments can be detected at the low collision energies involved and this makes an unambiguous detection of multifragmentation impossible. The situation is different for very high collision energies in the MeV range where there is also a high probability for detection of neutral fragments. Coincidence experiments have been used to investigate cluster-cluster collisions under such high collision energies, using C_{60} as the target for H_{25}^{+} projectile ions [32,33]. The observed power law in the fragment distributions with increasing collision energy along with the increasing multplicity is a strong indication of a liquid-gas phase transition such as discussed by Bonasera and co-workers [34].

2. CHARGE TRANSFER

The study of charge transfer in collisions where fullerenes are both the projectile ion and the target has been carried out for collision energies ranging from thermal to 100 keV and for projectile charges ranging from +1 to +6 [35-40]. These collision systems have been of interest for exploring the application of simple phenomenological charge transfer models, borrowed from atomic physics, such as the over-the-barrier model (Chap. 3.4). The results also help to provide a detailed understanding of the electronic properties of the isolated fullerenes and the conditions for electrical contact that may be highly relevant in future for producing nanoscale electrical devices based on fullerenes and carbon nanotubes.

The measured cross sections for collisions between fullerene ions and neutral fullerene targets at a laboratory collision of 100 keV are given in Table 9.3 [36]. Here $\sigma_{qq'}$ refers to collisions with a q+ charged projectile ion that result in a q'+ charged product. The single capture cross sections (σ_{21} and σ_{32}) are very similar in magnitude. This can be explained with the simple classical over-the-barrier argument, in this case using the "movable hole" picture (Chap.3.4.1). In this case, the high polarisability of the neutral fullerene target will tend to orient the initially delocalised holes on the projectile ion so that one positive charge will be at the minimum distance to the target. At the same time, the target will itself become polarised so that if an electron is transferred, it will start out from the point on the target that is closest to the projectile. In a first approximation, the electron transfer from the target to the projectile can be modeled as a singly-charged ion-atom collision, with an ion-atom separation equal to the minimum distance between the fullerenes. The classical barrier argument then yields a critical separation between the fullerenes, R_c = 11a_o. The cross-section for single-electron capture then becomes:

$$\sigma_{a(a-1)} = \frac{q}{q+1}\pi\left[(2r_s + R_C)^2 - (2r_S)^2\right] \tag{9.2.1}$$

where r_S is the fullerene radius. This gives values of σ_{21} = 280 Å^2 or 240 Å^2 calculated for r_S = 8.2 a_0 and 6.7 a_o, corresponding to the geometrical cage radius and to the value obtained from the polarisability, respectively. The corresponding values for σ_{32} are 310 Å^2 and 270 Å^2. The agreement with experiment is really very good.

To explain the double-electron capture by a triply-charged projectile one can consider the relation between the geometry of the two positive charges left on the projectile after capture of the first electron and a positive charge now on the target, Fig. 9.18. Since the two ions are close to one another there are two different possible configurations. The first is that the positive charge is repelled by the projectile and moves to the far side of the fullerene cage. This is the most likely situation for large impact parameters and it does not lead to any further electron capture. The second possibility, which is only stable for small impact parameters, is where the three positive charges all fall on a straight line passing through the fullerene centres, Fig. 9.18. This is possible because there is a polarisation attraction on one of the positive charges of the projectile by the target. The competition between this attraction and the Coulomb repulsion from the target positive charge can be estimated to give a critical distance for further charge transfer of $R_C \approx 7a_0$. The classical over-the-barrier cross section is then

$$\sigma_{31} = \frac{1}{2}\pi\left[(2r_S + R_{C2})^2 - (2r_S)^2\right] \tag{9.2.2}$$

which gives 100 Å^2 or 120 Å^2, depending on the value used for r_S, again in reasonably good agreement with the experimental values [36].

Table 9-3. Experimental cross sections $\sigma_{qq'}$ for charge transfer in units of Å^2 for collisions between fullerenes at a laboratory collision energy of 100 keV. q and q' refer to the projectile ion charge before and after collision. Data from [36].

Target	Projectile	σ_{21}	σ_{32}	σ_{31}	σ_{1-1}
C_{60}	C_{60}	260±80			0.13±0.04
C_{60}	C_{70}	210±65	210±65	63±20	

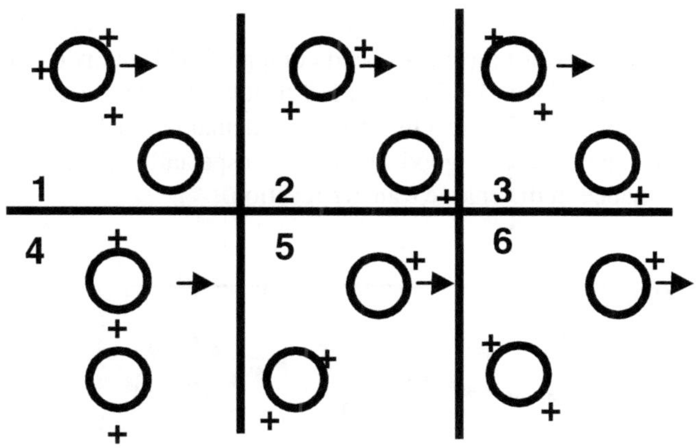

Figure 9.18. Schematic picture of double electron transfer to a triply charged fullerene form a neutral fullerene. Adapted from [36].

The cross section for double electron transfer to a singly-charged projectile to form the anion is orders of magnitude smaller than the other charge transfer reactions and cannot be estimated with the simple classical over-the-barrier arguments.

Studies of collisions between a neutral C_{60} target and charged projectile ions, C_{60}^{q+} (q = 2-5) at collision energies of 2q keV confirmed the basic picture for the charge transfer mechanism [40]. An extension of the simple model of Fig. 9.18 to higher projectile charges predicts that the maximum observable target charge after collision with projectiles of q = 1,2,3,5 should be 1,1,2,2. This is in perfect agreement with the experimental results. The results also show that electrical contact between two isolated, charged C_{60} is already established at a centre-to-centre distance in excess of 20 a_0 and that the charge re-distribution must occur on a time scale shorter than 10^{-14} to 10^{-15} s.

The values shown in Table 9.3 are for a relatively high collision energy. However, the absolute values of the cross sections do not change very much with collision energy. Cross sections are plotted as a function of collision energy for much lower energy collisions in Fig. 9.19 and 9.20. Fig. 9.19 shows the measured cross sections for C_{60}^{q+} + C_{60} charge transfer collisions for q = 2,3. The value of the cross sections for single and double electron transfer to a projectile with q = 3 are ca. 300-400 Å^2 and 100-200 Å^2, respectively in the collision energy range of a few keV [38], in very good agreement with the simple model discussed above. Similarly good agreement is obtained from single electron transfer collisions involving doubly charged projectiles where the measured cross section is scattered around 100-300 Å^2, to be compared with the model estimate of 240-280 Å^2. The low energy collisions have also been compared with the two-state Rapp and Francis model (Chap. 3.4.2), modified to take the size of the fullerene cage into account [37]. The results of these calculations are also shown in Fig. 9.19 and again there is good agreement, especially if one assumes a small off-resonance in the charge transfer reaction.

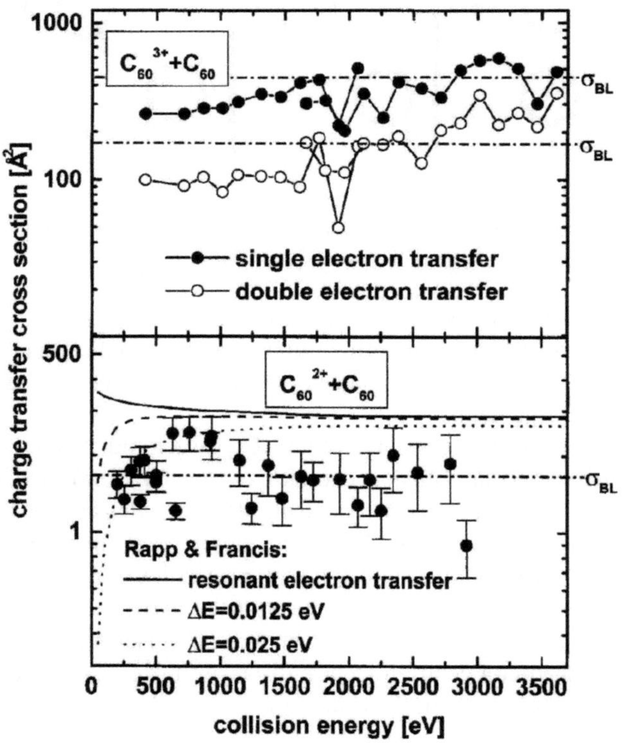

Figure 9.19. Absolute charge transfer cross sections for C_{60}^{q+}+C_{60} collisions as a function of centre of mass collision energy. Top: $q = 3$ [38] single and double electron transfer. Bottom: $q = 2$ [37] single electron transfer.

In all of these studies it is assumed that charge transfer occurs predominantly for grazing collisions and the experiments generally discriminate against low impact parameter collisions that would lead to scattering as well as transfer of excitation energy to vibrational degrees of freedom, leading ultimately to fragmentation. This is a reasonable assumption for high energy collisions and collisions with multiply-charged projectile ions. One has to be more careful with low energy collisions using a singly-charged projectile ion. Results of the measurements of charge transfer cross sections for single-electron transfer in collisions between C_{60}^+ + C_{60} are shown in Fig. 9.20. The original publication only presented results for products detected within a laboratory scattering angle range of $0 \pm 0.6^{\circ}$ [39]. Later investigations of differential cross sections showed that scattering outside this range was in fact significant for fragmentationless charge transfer reactions with singly charged projectiles and collision energies in the range of a few hundred eV to a few keV [10]. The original data is shown in Fig. 9.20 along with the cross sections obtained from integrating over the scattering angle dependence of the charge transfer signal as a function of collision energy. Clearly, the earlier cross sections were underestimated by a factor slightly larger than 2 for the investigated collision energy range. The correction brings the cross section value into very good agreement with the other single-electron transfer cross sections and the

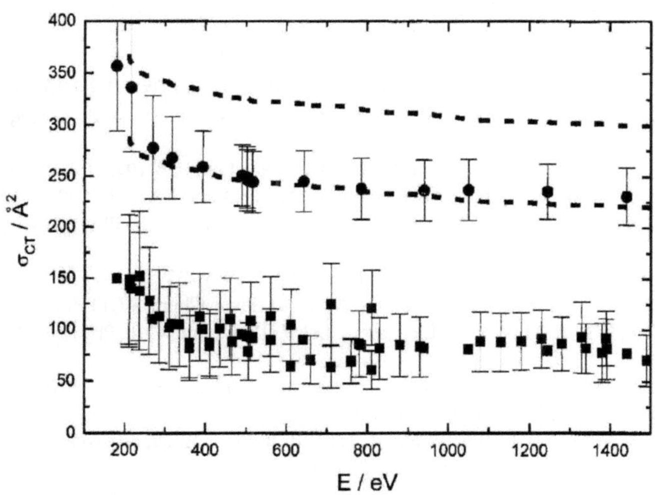

Figure 9.20. Total resonant charge transfer cross section for C_{60}^++C_{60} collisions. Squares: original data not accounting for effects of scattering [39]. Circles: data corrected for scattering [10]. Dotted lines: upper and lower limits estimated from the Rapp and Francis model [39].

prediction of the over-the-barrier model (210-240 Å^2). It is also in good agreement both in terms of magnitude and in terms of energy dependence for low energies with the prediction of the Rapp and Francis model for resonant charge transfer [39].

REFERENCES

1 R. Schmidt and H. O. Lutz, Phys. Rev. A **45,** (1992) 7981.
2 R. Mitzner, B. Winter, C. Kusch, E. E. B. Campbell, and I. V. Hertel, Z. Phys. D **37,** (1996) 89.
3 C. Yeretzian, K. Hansen, F. Diederich, and R. L. Whetten, Nature **359,** (1992) 44.
4 A. M. Rao, P. Zhou, K.-A. Wang, G. T. Hager, J. M. Holden, Y. Wang, W.-T. Lee, X.-X. Bi, P. C. Eklund, D. S. Cornett, M. A. Duncan, and I. J. Amster, Science **259,** (1993) 955.
5 E. E. B. Campbell, V. Schyja, R. Ehlich, and I. V. Hertel, Phys. Rev. Lett. **70,** (1993) 263.
6 F. Rohmund, E. E. B. Campbell, O. Knospe, G. Seifert, and R. Schmidt, Phys. Rev. Lett. **76,** (1996) 3289.
7 F. Rohmund, A. V. Glotov, K. Hansen, and E. E. B. Campbell, J. Phys. B. **29,** (1996) 5143.
8 A. V. Glotov, O. Knospe, R. Schmidt, and E. E. B. Campbell, Eur. J. Phys. D **16,** (2001) 333.
9 A. V. Glotov and E. E. B. Campbell, Phys. Rev. A **62,** (2000) 033202.
10 A. V. Glotov and E. E. B. Campbell, Chem. Phys. Lett. **327,** (2000) 61.
11 A. V. Glotov, PhD Thesis, Göteborg University, Sweden, (2000).
12 O. Knospe and R. Schmidt, in *Theory of Atomic and Molecular Clusters with a Glimpse at Experiments*, edited by J. Jellinek (Springer Verlag, 1999), p. 111.
13 L. Liu, K. Chen, and Y. Li, Chin. Phys. Lett. **10,** (1993) 720.
14 D. H. Robertson, D. W. Brenner, and C. T. White, J. Phys. Chem. **99,** (1995) 15721.
15 D. H. Robertson, D. W. Brenner, and C. T. White, Mat. Res. Soc. Symp. Proc. **359,** (1995) 199.
16 R. Schmidt, J. Schulte, O. Knospe, and G.Seifert, Physics Letters A **194,** (1994) 101.
17 J. Schulte, O. Knospe, G. Seifert, and R. Schmidt, Physics Letters A **198,** (1995) 51.
18 Y. Xia, Y. Xing, C. Tan, L. Mei, and H. Yang, Nucl. Instr. Meth. Phys. Res. B **111,** (1996) 41.
19 Y. Xia, Y. Xing, C. Tan, and L. Mei, Phys. Rev. B **53,** (1996) 13871.
20 B. L. Zhang, C. Z. Wang, C. T. Chan, and K. M. Ho, J. Phys. Chem. **97,** (1993) 3134.
21 E. E. B. Campbell, A. Glotov, A. Lassesson, and R. D. Levine, C.R. Physique **3,** (2002) 341.
22 R. D. Levine and R. B. Bernstein, *Molecular Reaction Dynamics and Chemical Reactivity* (Oxford University Press, New York, Oxford, 1987).
23 O. Knospe, A. V. Glotov, G. Seifert, and R. Schmidt, J. Phys. B. **29,** (1996) 5163.
24 R. Ehlich, PhD Thesis (Universität Freiburg), (1994).
25 D. L. Strout, R. L. Murry, C. Xu, W. C. Eckhoff, G. K. Odom, and G. E. Scuseria, Chem. Phys. Lett. **214,** (1993) 576.
26 F. Rohmund, PhD Thesis, FU Berlin, (1997).
27 S. Petrie and D. K. Bohme, Nature **365,** (1993) 426.
28 T. Lill, F. Lacher, H.-G. Busmann, and I. V. Hertel, Phys. Rev. Lett. **71,** (1993) 3383.
29 E. E. B. Campbell, T. Raz, and R. D. Levine, Chem. Phys. Lett. **253,** (1996) 261.
30 G. Seifert, R. Gutierrez, and R. Schmidt, Phys. Lett. A **211,** (1996) 357.

[31] J. Schulte, Int. J. Mass Spec. Ion Proc. **145,** (1995) 203.
[32] B. Farizon, M. Farizon, and M. J. Gaillard, Int. J. Mass Spectr. **192,** (1999) 259.
[33] B. Farizon, M. Farizon, M. J. Gaillard, F. Gobet, C. Guillermier, M. Carré, J. P. Buchet, P. Scheier, and T. D. Märk, Eur. Phys. J. D **5,** (1999) 5.
[34] M. Belkacem, V. Latora, and A. Bonasera, Phys. Rev. C **52,** (1995) 271.
[35] D. K. Bohme, Int. Rev. Phys. Chem. **13,** (1994) 163.
[36] H. Shen, P. Hvelplund, D. Mathur, A. Barany, H. Cederquist, and N. Selberg, Phys. Rev. A **52,** (1995) 3847.
[37] F. Rohmund and E. E. B. Campbell, Chem. Phys. Lett. **245,** (1995) 237.
[38] F. Rohmund and E. E. B. Campbell, Z. Phys. D **40,** (1997) 399.
[39] F. Rohmund and E. E. B. Campbell, J. Phys. B **29,** (1997) 5143.
[40] H. Cederquist, P. Hvelplund, H. Lebius, H. T. Schmidt, S. Tomita, and B. A. Huber, Phys. Rev. A **63,** (2001) 025201.

Chapter 10

SURFACE COLLISIONS

"I hurried to complete the paper sphere...When I dropped it on the floor, it actually bounced. Surely, this had to be the correct shape." This quotation from Rick Smalley [1] comes from a description of the search for the structure of C_{60} and can perhaps be regarded as the first report on a surface collision experiment with the molecule! Fullerene surface collisions received a considerable amount of interest after initial reports of the extremely high resilience to fragmentation of C_{60}^+ at relatively high impact energies [2]. The observation was also in agreement with the predictions of classical MD simulations for collisions between a fullerene and a diamond surface [3]. Many of the phenomena observed in gas phase collisions have their counterpart in surface collisions. A schematic summary of the processes that have been investigated is shown in Fig. 10.1. In this chapter we will review the field of fullerene-surface collisions. The emphasis will be on the basic scattering dynamics but we will also touch upon possible applications that exploit the unique nature of the fullerene collisions.

1. FRAGMENTATION AND SCATTERING

1.1 Fragmentation

Most of the experiments that will be described in this chapter have been carried out in apparatus that is similar to that shown in Fig, 2.17. A beam of fullerene ions is produced (normally, laser desorption/ionisation has been used to produce the fullerene cations for the surface scattering experiments),

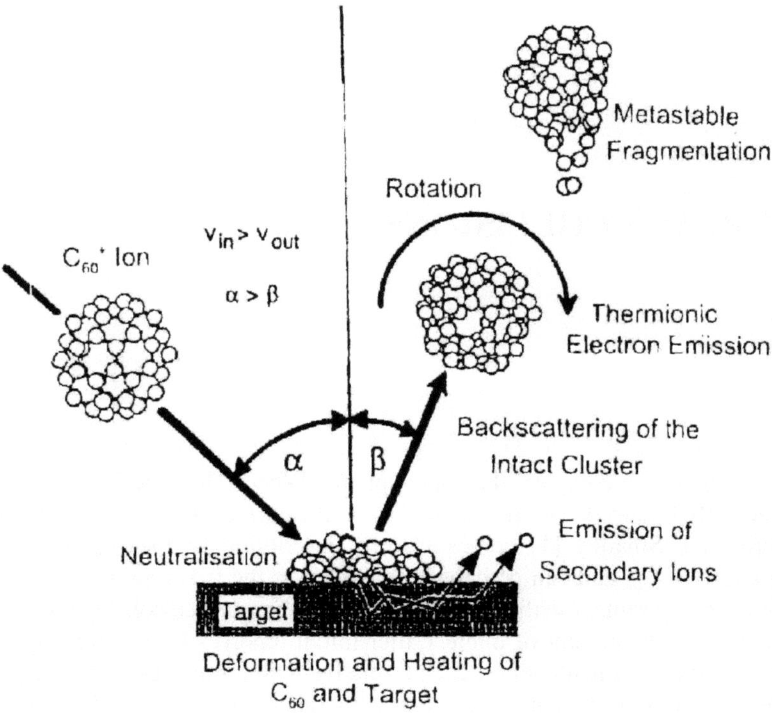

Figure 10.1. Schematic summary of the processes observed in collisions of C_{60}+ ions with surfaces.

mass-selected, accelerated or decelerated if necessary and collided with a surface (silicon, graphite and diamond have been popular choices). The scattered fullerene ions and their fragments are then detected by time-of-flight mass spectrometry. Early experiments reported that the surface collisions in the impact energy range of 50 - 500 eV were highly inelastic, resulting in a nearly constant kinetic energy of the scattered ions of 5-10 eV [2, 4-6]. A value of 25-35 % energy transfer to internal degrees of freedom was assumed based on classical MD simulations for collisions with diamond [3]. It was later shown [7] that the kinetic energy of the scattered fullerenes as well as their internal energy after the surface collision was strongly dependent on the cleanliness of the surface. The graphite targets used in the early experiments had a residual hydrocarbon coating that affected the collision dynamics. Time-of-flight mass spectra of the products of collisions

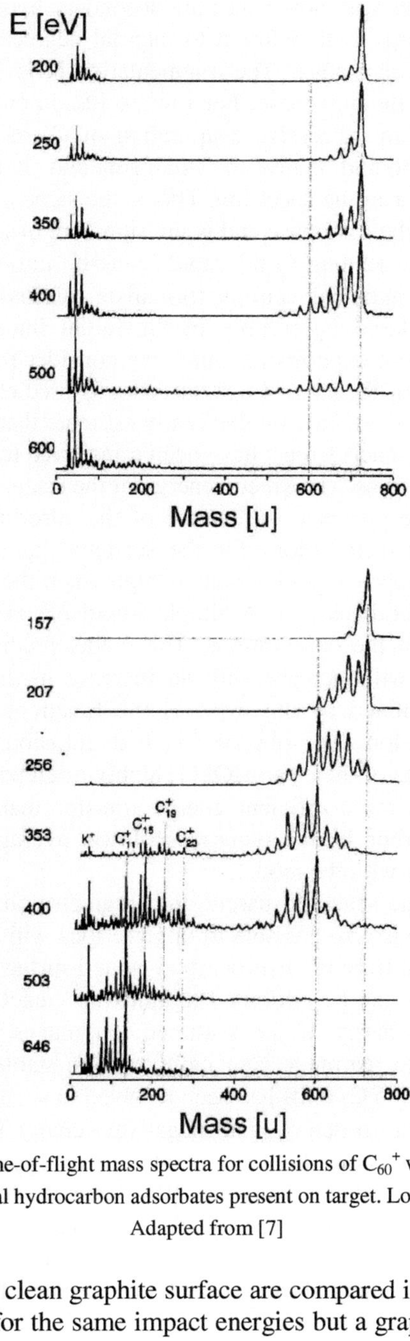

Figure 10.2. Time-of-flight mass spectra for collisions of C_{60}^+ with a graphite target. Upper: with residual hydrocarbon adsorbates present on target. Lower: clean graphite. Adapted from [7]

between C_{60}^+ and a clean graphite surface are compared in Fig. 10.2 with the products obtained for the same impact energies but a graphite target that has hydrocarbon adsorbates present. The small masses present at short flight times from the target with adsorbates are secondary hydrocarbon ions. It is clear that more energy is transferred to internal degrees of freedom in the collisions with a clean surface. The fragmentation behaviour is very similar to that observed in the gas phase. For low excitation energies one sees the typical pattern due to successive evaporation of C_2 molecules. At impact energies between 256 and 353 eV the small ring and chain geometry carbon fragment ions appear in the spectrum. This is the same as for collisional (or laser) excitation in the gas phase and is the signature of a phase transition in the excited fullerene leading to the rapid break-up into small fragments as modelled with the maximal entropy formalism, discussed in Chap. 3.2.2. There is a rather large uncertainty in the initial internal energy of the fullerene ions in these experiments but if we consider the threshold energy for the appearance of the small fragment ions, as predicted by the maximal entropy model (ca. 65 eV [8]) we can easily estimate that approximately 20-25% of the collision energy must have been transferred to internal degrees of freedom in the collisions. The kinetic energy of the scattered fullerene ions is also affected by the presence or absence of the adsorbate layer. Fig. 10.3 shows the velocity distributions for the scattered parent ions. The post-impact velocity is approximately twice as high when the graphite target has no hydrocarbon adsorbates [7]. A simple impulsive excitation model can qualitatively explain the observations. The model predicts that the energy transfer efficiency will increase with an increase in the effective surface mass that is determined by the mass of the functional group hit by the incident molecular ion. The observation that collisions with the densely packed carbon atoms of the clean HOPG (highly oriented pyrolytic graphite) surface result in a more efficient energy transfer than collisions with a disordered hydrocarbon layer having much lower average atomic mass and density is consistent with the model.

In addition to the scattered parent and fragment ions and the sputtered hydrocarbon ions it is also possible to observe ions with masses larger than C_{60}^+ when scattering from the hydrocarbon-coated surface [7] (barely visible in Fig. 10.2). This can be attributed to "pick-up" reactions at the surface. The high internal energy of the scattered ion makes it possible for the collision complex to rearrange ("anneal") to form stable higher fullerenes. The addition of up to 5 C_2 units has been resolved in such experiments.

It is also possible to detect small negatively charged carbon clusters as the products of collisions between positively charged C_{60}^+ and various targets (e.g. graphite, silicon, tantalum) [9]. Some results are shown in Fig. 10.4 as a function of collision energy. The positive and negative integrated fragment ion yields are shown in Fig. 10.5 as a function of collision energy. The small negatively charged carbon clusters are neutral fragments from the

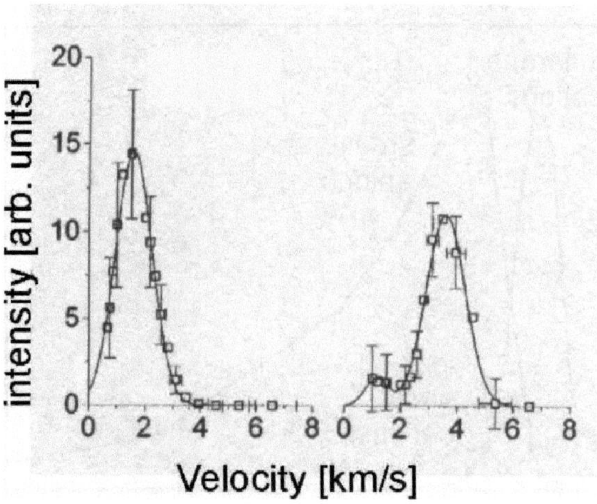

Figure 10.3. Measured velocity distributions for C_{60}^+ scattered from graphite at an impact energy of 200 eV. Left: target with residual hydrocarbon adsorbates. Right: clean target. Adapted from [7].

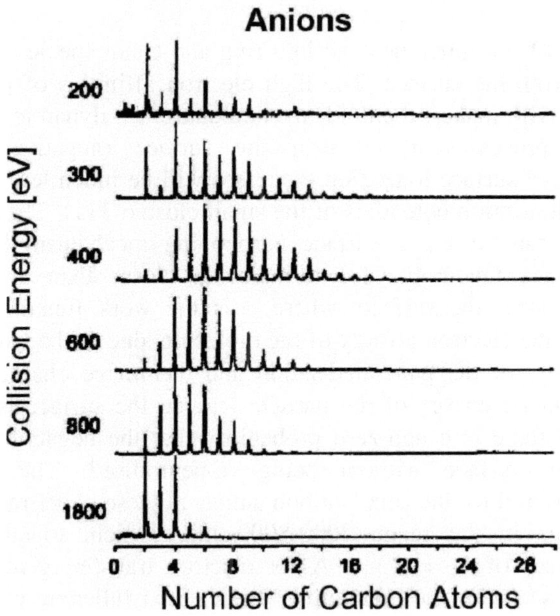

Figure 10.4. Small carbon anions observed from collisions between C_{60}^+ and a graphite target for different impact energies. The signal is due to small neutral fragments "picking-up" an electron from the surface. Adapted from [9].

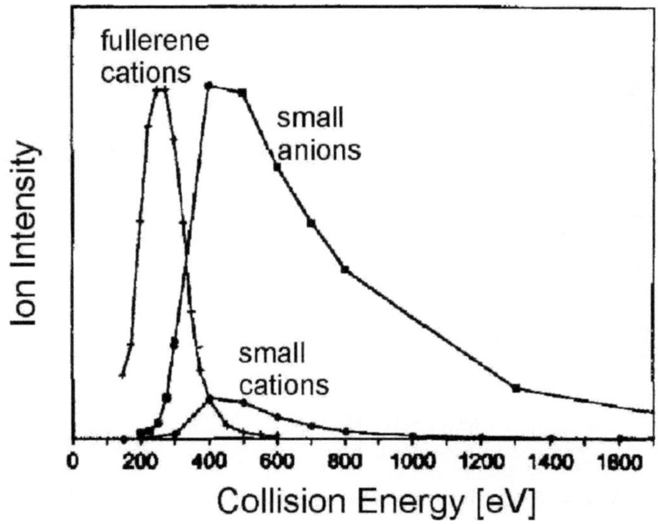

Figure 10.5. Integrated intensities of detected fragment species as a function of C_{60}^{+} impact energy on graphite. Adapted from [9].

prompt break-up of the fullerene cage into ring and chain species that pick up an electron from the surface. The high electron affinities of the small carbon clusters [10] makes them ideal candidates for dynamic electron transfer/pick-up processes at or near the surface (negative surface ionisation). Positive surface ionisation processes will be much less efficient due to the large ionisation potentials of the small clusters [11]. The electron transfer process can be via a surface harpooning mechanism [12]. An electron transfer is favourable for a molecule closer than a distance $R_C = e^2 / 4(\phi - E_a)$ from the surface, where ϕ is the work function of the surface and E_a is the electron affinity of the molecule, due to the Coulombic stabilisation energy of the generated anion and the image charge on the surface. If the kinetic energy of the particle leaving the surface is greater than ϕ - E_a then there is a non-zero probability that the negative ion can escape from the surface without being re-neutralised. The velocity distributions measured for the small carbon anions in these experiments give average velocities in the range 2000-8000 ms^{-1} which, together with maximum distances of R_C = 2 - 6 Å for electron transfer, gives a sub-picosecond time scale for the break-up of the excited fullerene cage. This agrees very well with MD simulations of highly excited fullerenes [13-15].
The formation of negative parent ions can be seen in experiments involving hyperthermal collisions of neutral fullerene beams on surfaces [16].

Tight-binding MD simulations [15] of C_{60} collisions on a diamond surface indicated the presence of three different collision regimes. For low collision energies, $E \leq 120$ eV, the molecule bounces off the surface without forming any bonds with the surface atoms and recovers its original shape after severe distortions. In the intermediate regime, $120 \leq E \leq 240$ eV, C_{60} forms covalent bonds with the surface. The molecule can either be adsorbed or can leave the substrate with defects in its original structure, leading to metastable evaporation of C_2 units. Similar results have been found for simulations using a semi-empirical density functional approach [17], Fig. 10.6. Finally, for $E \geq 240$ eV, the molecule breaks after impact with the substrate, forming large ring and chain fragments. Pieces of the broken cage can form stable bonds with the substrate which can induce a local surface

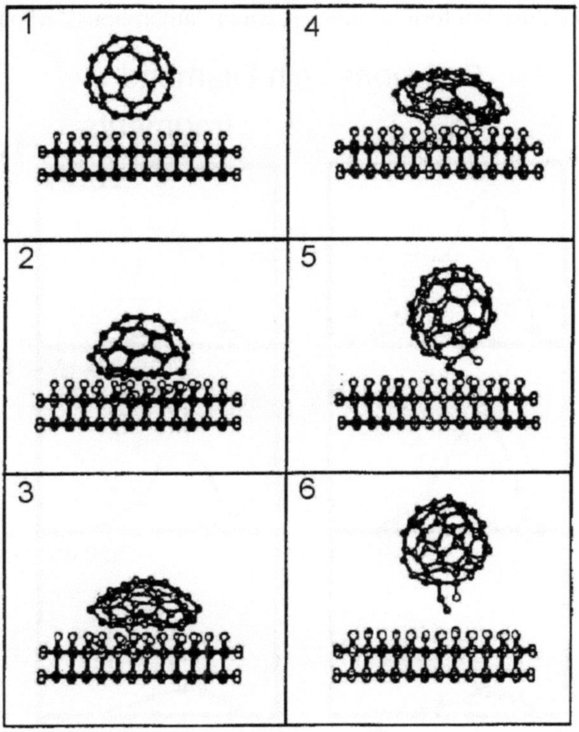

Figure 10.6. MD simulations of a C_{60} colliding with a diamond surface at an impact energy of 150 eV and an angle of incidence of 20°. Adapted from [18]. See also [17] for details of the calculation.

reconstruction. The calculations also showed that the kinetic energy of the fullerene centre of mass, when it bounces off the surface shows a clear dependence on the incident energy in the low collision energy regime. However, in the intermediate regime, the centre-of-mass energy after impact depends only weakly on the impact energy [15]. The calculations are in good qualitative agreement with measurements of C_{60}^{+} scattered on diamond [18] in the collision energy range of 100-1500 eV. Velocity distributions for collision energies in the intermediate range are shown in Fig. 10.7. All distributions measured in this range (220 - 455 eV) show a velocity distribution that peaks at approximately 2000 ms^{-1} and is thus independent of the impact energy. The intensity of scattered ions decreases strongly for impact energies greater than 300 eV and for energies beyond 500 eV, no scattered ions could be observed at all. This can be explained by complete destruction and reaction of the fullerene with the surface atoms. Such a process has been used to form a stable, resistant amorphous carbon coating

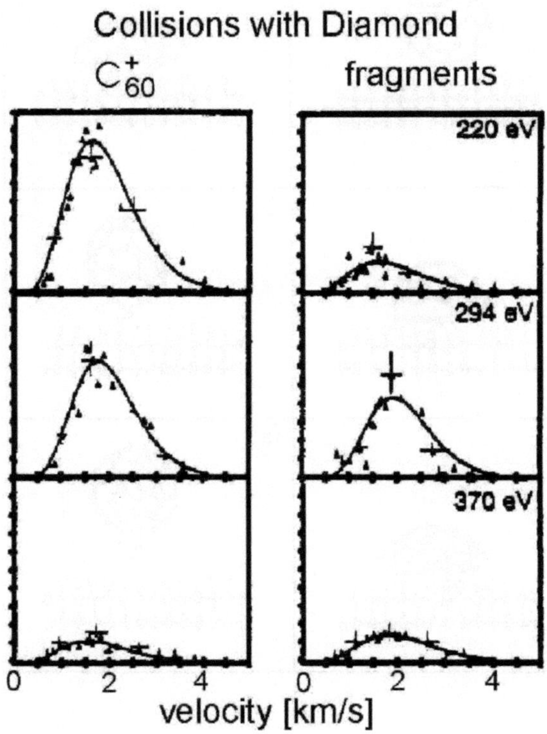

Figure 10.7. Velocity distributions of scattered C_{60}^{+} (left) and fragment ions (right) for collisions with a diamond (111) surface at different impact energies and an angle of incidence of 20°. Adapted from [18].

on air-sensitive fullerite films [19].

Lower energy collisions were investigated by decreasing the angle of incidence of the fullerene beam on the surface thus decreasing the normal component of the initial velocity. The experimental geometry is shown in Fig. 10.8. For these experiments it was not possible to rotate the time-of-flight mass spectrometer, as in Fig. 2.17, instead two fixed detector positions were used and the target could be rotated to vary the angle of incidence. The dependence of the final normal velocity to the initial normal velocity is plotted in Fig. 10.9 and shows an initial increase in the low collision velocity regime, reaching a plateau for velocities corresponding to kinetic energies beyond 120 - 140 eV, in good agreement with the MD simulations [15].

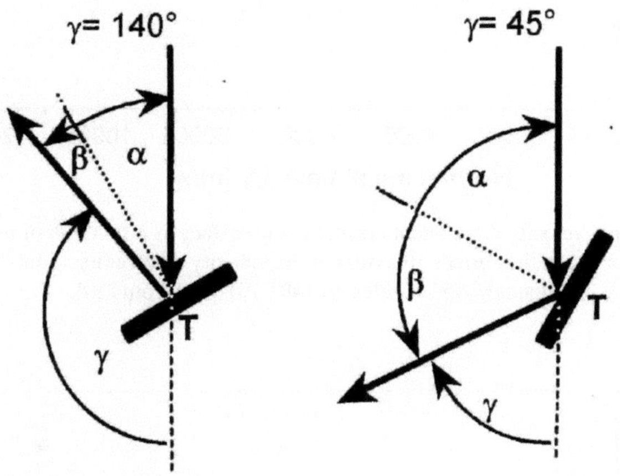

Figure 10.8. Experimental geometry for the experiments described in [18, 20]. The positions of two time-of-flight detectors are fixed corresponding to the shown γ angles. The angle of incidence can be varied by rotating the target.

A similar dependence of the final kinetic energy with respect to the initial kinetic energy was shown in collisions between a hyperthermal neutral C_{60} beam and a carbon-passivated nickel surface for impact energies in the range 10-50 eV [21]. The kinetic energy of the scattered molecules is plotted as a function of the initial kinetic energy for two angles of incidence in Fig. 10.10. A linear dependence of the scattered energy on the initial energy is clearly seen from the low impact energy data.

1.2 Scattering

In the above section we concentrated on the component of the initial kinetic energy of the projectile normal to the surface. Here we focus on the tangential component. The behaviour of the tangential velocity component is

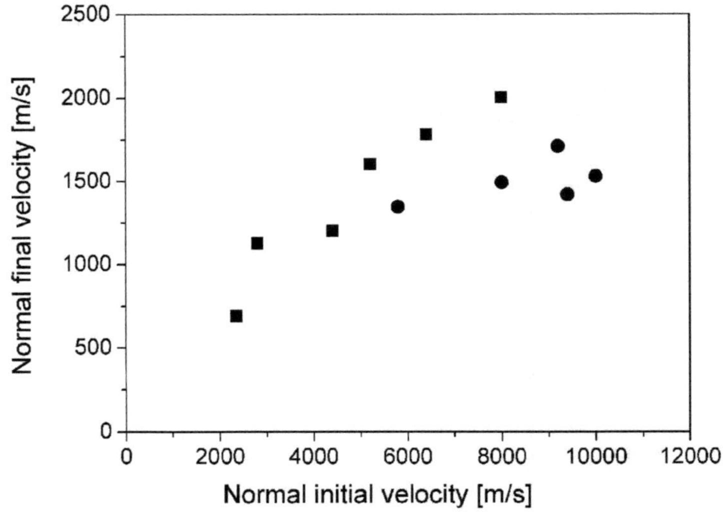

Figure 10.9. Final velocity component normal to the surface as a function of initial velocity component normal to the surface (maxima of the velocity distributions) for C_{60}^{+} impact. Squares γ=45°. Circles: γ=140°. Adapted from [20].

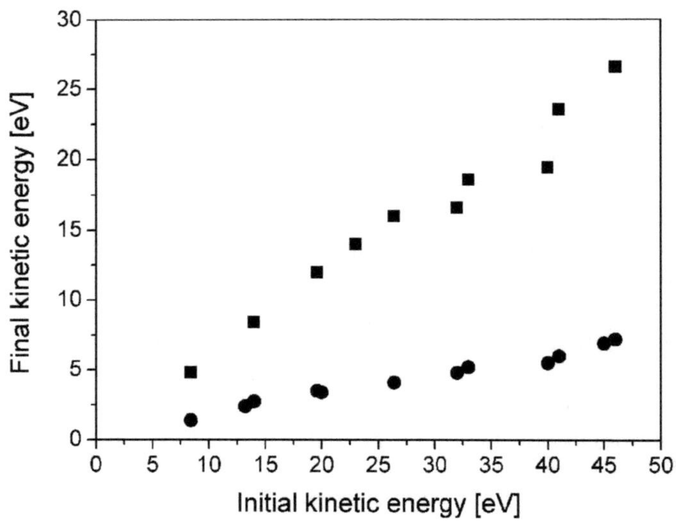

Figure 10.10. Relationship between initial and final kinetic energy of C_{60} colliding on a carbon-passivated Ni surface. Squares: incident angle of 61°. Circles: incident angle of 15.7°. Data from [21].

closely related to the rotational angular momentum transferred to the molecule. The fullerene behaves very much like a rubber ball picking up top-spin in a collision with a solid surface [20-23].

The collision of a C_{60} molecule with a solid surface is illustrated schematically in Fig. 10.11. We assume that the C_{60} hollow sphere touches the surface at any given time at just one point P only and in such a manner that no torque is applied with respect to that contact point. Assuming the absence of initial rotation, conservation of angular momentum gives a simple kinematic constraint

$$amu_t = amv_t + L \qquad (10.1)$$

where u_t and v_t are the initial and final tangential velocity, respectively, a is the distance of closest approach, m the fullerene mass and $L = I\Omega$ the angular moment of the ball after the collision. $I = qma^2$ is the moment of inertia at the time of closest approach with the numerical factor $q = 2/3$ for an undeformed hollow sphere. This can be taken as a good approximation for low energies. As the energy increases, the spherical molecule becomes deformed under the influence of normal forces into an ellipsoid as shown in Fig. 10.11. Equation (10.1) can be written as

$$v_t = u_t - qa\Omega \qquad (10.2)$$

which gives the relationship between the tangential velocities and the velocity of rotation $a\Omega$ at the surface of the fullerene. Equation (10.2) predicts that the final tangential velocity, v_t, will be smaller than the initial velocity component parallel to the surface, u_t. Also, the loss of tangential velocity during the collision is predicted to increase for larger factors, q, i.e. with increasing deformation of the sphere. An upper bound for $a\Omega$ is given for complete sticking of the contact point to the surface, $a\Omega = u_t$. Then one obtains $v_t = u_t\ (1-q)$, giving $v_t = 1/3$ for an undeformed sphere and a decreasing v_t/u_t ratio at higher impact energies where q becomes larger. Figure 10.12. shows the experimentally determined v_t/u_t ratio for different angles of incidence and impact energies [20]. The experimental values are very close to 1/3 for low impact energies and start to deviate significantly to lower values for energies greater than 800 eV (corresponding to a normal impact energy of 120 eV), in very good agreement with the simple rolling-ball model. A similar model was applied to the experiments using a hyperthermal neutral fullerene beam [21]. A deviation from the simple model was observed which could be attributed to the existence of translational slip for the low energy neutral collisions [21].

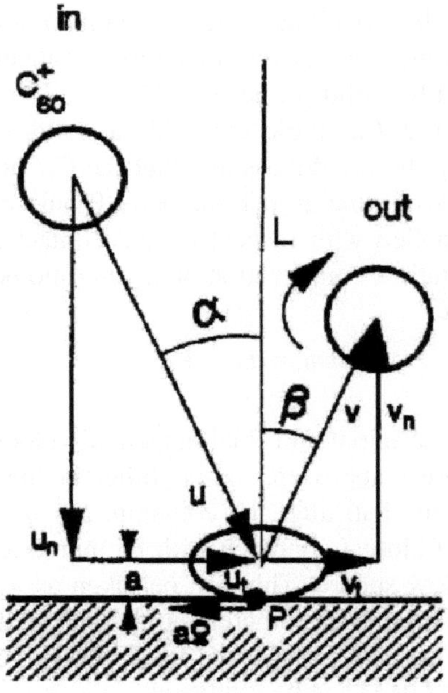

Figure10.11. Schematic picture of the scattering geometry illustrating the transfer of tangential velocity into rotational angular momentum of the C_{60}^{+} during the surface collision. Adapted from [22].

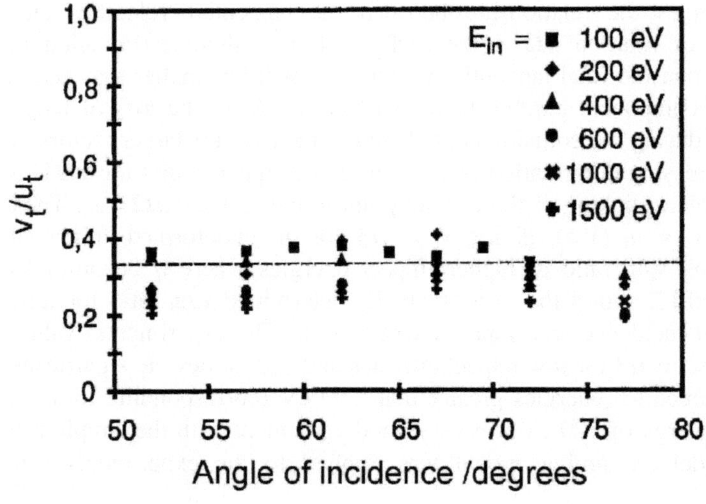

Figure 10.12. Ratio of final and initial tangential velocities as a function of the angle of incidence for different impact energies. The data are clusteres around $v_t = u_t/3$ as expected from the simple rolling ball picture. Adapted from [20].

1.3 Collisions with Fullerite Films

Fullerite films (epitaxial films of fullerene molecules) have three properties that are important for the C_{60}-surface scattering process. Firstly, the projectile and the target molecules in the surface have the same mass. Secondly the surface corrugation is large and its period matches the diameter of the impinging fullerene. Thirdly, the surface consists of only weakly bound molecules. These three characteristics are responsible for the different results obtained from fullerene-fullerite collisions [18] compared to scattering of fullerenes from more conventional surfaces, discussed above.

Fig. 10.13 shows a typical mass spectrum of ions after C_{60}^+ impact on a fullerite film [24]. In this particular example the initial kinetic energy of the projectile beam was 275 eV and the incident angle was 25 ^{o} with respect to the surface normal. The products from colliding with a graphite surface under identical experimental conditions are shown in the bottom part of the figure. Practically no fragmentation is observed in the fullerite scattering case but there is a signal corresponding to masses much larger than the projectile ion. This is due to a molecular fusion reaction between the projectile fullerene ion and a target fullerene on the surface. The reaction is very similar to the gas phase fusion discussed in Chap. 9. One can clearly see a typical fullerene fragmentation pattern for the larger molecules corresponding to the loss of multiple C_2 units indicating that there is sufficient internal energy present for a rapid rearrangement of the carbon bonds to produce a metastable C_{120}^+ isomer rather than a C_{60} dimer. The collision energy range for which these fusion products are detected is also consistent with the gas phase measurements. The velocity distributions of the product ions in these experiments also support this interpretation. The projectile ion beam consisted of mass-selected C_{60}^+ so that the detected C_{70}^+ ions must have been sputtered from the substrate (the fullerite film contained 10% C_{70}) [24]. There is a low velocity component of the C_{60}^+ ions that matches the C_{70}^+ distribution and that is also attributed to sputtering of the substrate material. The fusion products are significantly faster than the sputtered molecules and the velocity distribution is consistent with the formation of large fullerenes via fusion of the C_{60}^+ projectile ion with a surface fullerene in a direct reaction mechanism of the Eley-Rideal type [25].

Another interesting observation in fullerene-fullerite experiments is the occurrence of rainbow scattering. The rainbow effect, which can be observed in many kinds of particle scattering studies, causes enhanced scattering intensities at certain scattering angles, the so-called rainbow angles. It is named after the atmospheric phenomenon of rainbow arcs in the sky. In general, all rainbows are the result of two counter-acting effects. For atmospheric rainbows, it is the interaction of the reflection and refraction of

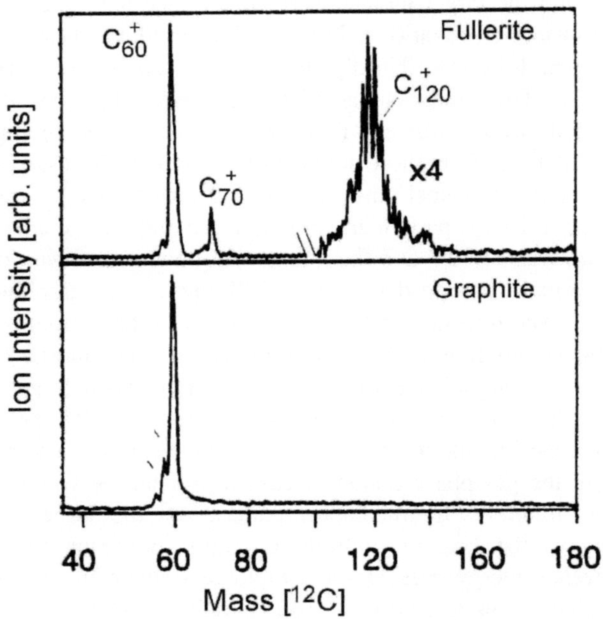

Figure 10.13. Time-of-flight mass spectra of the products of collisions between C_{60}^+ and (a) a fullerene film (b) graphite. Collision energy 275 eV and angle of incidence of 25°. Adapted from [24].

sunlight in water droplets that causes a minimum in the deflection angle. In atom-atom scattering it is the combination of the attractive and repulsive parts of the interaction potential. The angle of deflection changes from negative to positive as the impact parameter decreases. The minimum value of the angle of deflection is called the rainbow angle. According to the expression for the classical differential cross section (eq. (3.1.28)), at the rainbow angle $d\chi/db = 0$ and the classical deflection function goes to infinity. This shows up as a clear maximum in the experimental differential cross section. In surface scattering, the periodic surface potential gives rise to two extrema in the deflection function. The simplest case is scattering from a "zig-zag" potential where the reflection angle depends on which side of the "zig-zag" the projectile hits, see Fig. 10.14(a). The situation for fullerenes is shown schematically in Fig. 10.14(b). The primary angle of incidence, α, can be varied in the experiments while the angle γ is fixed at

140°. Therefore the detected reflection angle is determined by $\beta = 180^\circ - (\gamma + \alpha)$. The scattering plane is orthogonal to the fullerite surface and intersects it in the <110> direction [24]. The nearest neighbour distance in the C_{60} lattice is 10.02 Å in this direction. If the surface C_{60} molecules

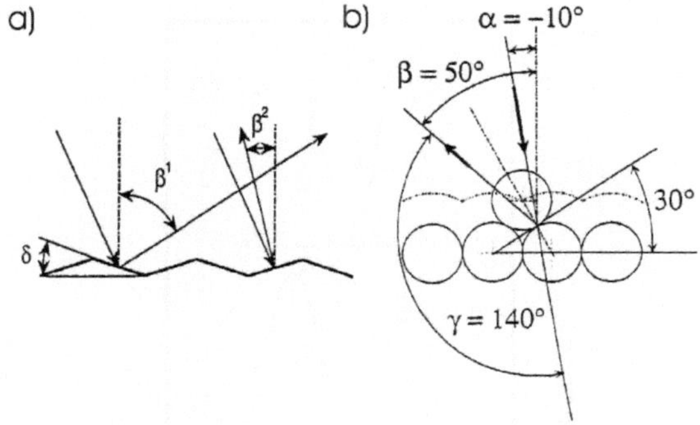

Figure 10.14. a) Illustration of specular reflection from a "zig-zag" potential. The particles are scattered at two angles: β1 and β2. (b) Scattering geometry for interaction of C_{60}^+ with a <111> oriented C_{60} crystal. The dot-dashed line shows an effective interaction potential. Scattering from this point under the condition $\alpha + \beta = 40^\circ$ results in enhanced scattering intensity for $\alpha = -10^\circ$.

interact with the projectile fullerene ion as hard spheres then the interaction process can be viewed as reflection of a point-like particle from the potential barrier indicated by the dash-dot line in Fig. 10.14(b). The dependence of the scattering signal on the angle of incidence α for a fixed $\gamma = 140^\circ$ is shown in Fig. 10.15 for different final velocities (the impact energy was 275 eV) [24]. Three partially overlapping maxima are found at approximately $\alpha = -10^\circ$, 20° and 50° which shift slightly as the detected velocity increases. The effect is even more pronounced for lower impact energies [18]. The calculations shown on the same figure come from an extremely oversimplified classical model and consider elastic reflection of a point mass from a rigid wall consisting of overlapping spheres (Fig. 10.14(b)). The calculations using a cosine form of the potential do indeed reproduce the pronounced rainbow structure. The central maximum can be attributed to lattice dynamics effects, as indicated by classical MD simulations [18].

2. SPUTTERING

In the previous section it was pointed out that there is a strong signal from sputtered ions when a fullerene projectile ion impinges on a fullerite target [24]. This has also been observed for very high energy collisions of fullerene ions (20 MeV) on bio-molecular samples [26]. For the desorption/ionisation of trypsin molecules, the secondary ion emission yield was found to be about two orders of magnitude higher than with MeV atomic ions. This is believed to be a consequence of the very high energy density deposited via electronic stopping in the solid by the highly energetic

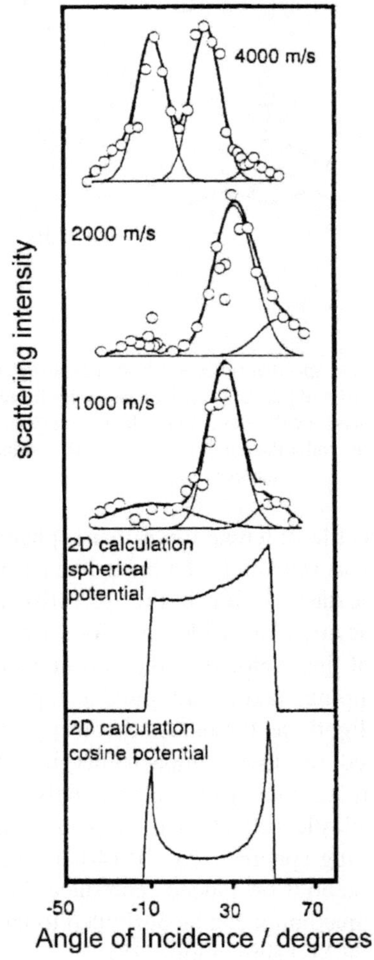

Figure 10.15. Angular distributions for C_{60}^+ scattered on a crystalline C_{60} films for different final velocities and an impact energy of 275 eV. The two lowest plots are the results of numerical 2D simulations with simplified potentials. From [24].

fullerene ions leading to the ejection of a large amount of matter after each ion impact. At lower impact energies fullerene ions are a very surface-sensitive probe and could potentially be used for SIMS (secondary ion mass spectrometry) applications [27]. The yield of ions per unit mass from organic surfaces was compared for small gold cluster ions and C_{60}^+ and found to be 4-5 times higher for the fullerenes than for gold projectiles of similar mass [28]. A TRIM calculations suggests that 20 keV C_{60}^+ deposits most of its energy within 30 Å of the surface whereas gold clusters of similar mass deposit their energy very much deeper in the solid (> 100 Å). A comparison of the positive secondary ion yields for Ga^+, Au^+, Au_3^+ and C_{60}^+ at 10 keV impact energy on poly(ethylene-terephthalate) is shown in Fig. 10.16 [27]. The advantage of using a fullerene ion beam is apparent. The damage cross sections for the organic materials under fullerene ion bombardment depend on their chemical and physical properties. High molecular weight polymers and monolayer films of molecular solids are damaged at approximately the same rate as under Ga^+ irradiation (an ion commonly used for SIMS applications). However, thick films of molecular solids show extremely low levels of damage and it appears that the molecules are being peeled off the surface with very little sub-surface damage. Generally, less fragmentation is seen with the fullerene projectile ions and there is a very large increase in the molecular ion yield compared to conventional projectile beams [27].

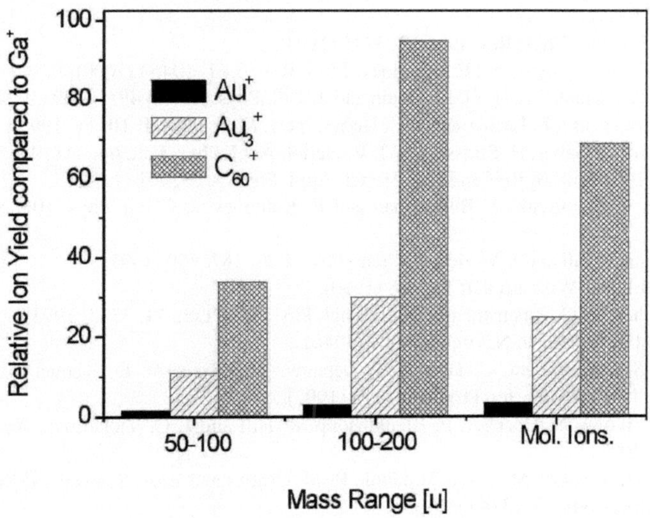

Figure 10.16. Comparison of positive secondary ion yield during analysis of PET using Au^+, Au_3^+ and C_{60}^+ at 10 keV impact energy. The intensities are normalised to the signal from Ga^+. Adapted from [27].

REFERENCES

1 R. E. Smalley, The Sciences **31**, 22 (1991).
2 R. D. Beck, P. S. John, M. M. Alvarez, F. Diederich and R. L. Whetten, J. Phys. Chem. **95**, 8402 (1991).
3 R. C. Mowrey, D. W. Brenner, B. I. Dunlap, J. W. Mintmire and C. T. White, J. Phys. Chem. **95**, 7138 (1991).
4 T. Lill, H. G. Busmann, B. Reif and I. V. Hertel, Surf. Sci. **272**, 146 (1992).
5 T. Lill, H. G. Busmann, B. Reif and I. V. Hertel, Chem. Phys. Lett. **187**, 459 (1991).
6 C. Yeretzian and R. L. Whetten, Z. Phys. D **24**, 199 (1992).
7 R. D. Beck, J. Rockenberger, P. Weis and M. M. Kappes, J. Chem. Phys. **104**, 3638 (1996).
8 A. Lassesson, A. Gromov, K. Mehlig, A. Taninaka, H. Shinohara and E. E. B. Campbell, J. Chem. Phys. **119**, 5591 (2003).
9 R. D. Beck, C. Warth, K. May and M. M. Kappes, Chem. Phys. Lett. **257**, 557 (1996).
10 A. Yang, T. K.J., M. J. Craycraft, J. Conceicao, C. L. Pettiette, O. Chesnovsky and R. E. Smalley, Chem. Phys. Lett. **144**, 431 (1988).
11 S. B. H. Bach and J. R. Eyler, J. Chem. Phys. **92**, 358 (1990).
12 J. W. Gadzuk and S. Holloway, J. Chem. Phys. **84**, 3502 (1986).
13 S. G. Kim and D. Tománek, Phys. Rev. Lett. **72**, 2418 (1994).
14 H. Hohmann, R. Ehlich, S. Furrer, O. Kittelmann, J. Ringling and E. E. B. Campbell, Z. Phys. D **33**, 143 (1995).
15 G. Galli and F. Mauri, Phys. Rev. Lett. **73**, 3471 (1994).
16 A. Bekkerman, B. Tsipinyuk and E. Kolodney, Phys. Rev. B **61**, 10463 (2000).
17 P. Blaudeck, T. Frauenheim, H.-G. Busmann and T. Lill, Phys. Rev. B **49**, 11409 (1994).
18 T. Lill, H.-G. Busmann, F. Lacher and I. V. Hertel, Int. J. Mod. Phys. B **10**, 11 (1996).
19 H.-G. Busmann, H. Gaber, H. Strasser and I. V. Hertel, Appl. Phys. Lett. **64**, 43 (1994).
20 T. Lill, H.-G. Busmann, B. Reif and I. V. Hertel, Appl. Phys. A **55**, 461 (1992).
21 A. Budrevich, B. Tsipinyuk, A. Bekkerman and E. Kolodney, J. Chem. Phys. **106**, 5771 (1997).
22 H.-G. Busmann, T. Lill and I. V. Hertel, Chem. Phys. Lett. **187**, 459 (1991).
23 W. Buerger, Bild der Wissenschaft **11**, 186 (1990).
24 T. Lill, F. Lacher, H.-G. Busmann and I. V. Hertel, Phys. Rev. Lett. **71**, 3383 (1993).
25 D. D. Eley and E. K. Rideal, Nature **146**, 401 (1940).
26 A. Brunelle, S. Della-Negra, C. Deprun, J. Depauw, P. Håkansson, D. Jacquet, Y. Le Beyec, *et al.*, Int. J. Mass Spec. Ion Proc. **164**, 193 (1997).
27 D. Weibel, S. Wong, N. Lockyer, P. Blenkinsopp, R. Hill and J. C. Vickerman, Analyt. Chem. **75**, 1754 (2003).
28 R. D. Harris, W. S. baker, M. J. van Stipdonk, R. M. Crooks and E. A. Schweikert, Rapid Commun. Mass Spectrom. **13**, 1374 (1999).

MIX
Papier aus verantwortungsvollen Quellen
Paper from responsible sources
FSC® C105338

If you have any concerns about our products,
you can contact us on
ProductSafety@springernature.com

In case Publisher is established outside the EU,
the EU authorized representative is:
Springer Nature Customer Service Center GmbH
Europaplatz 3, 69115 Heidelberg, Germany

Printed by Libri Plureos GmbH
in Hamburg, Germany